MATHEMATICS DEPARTMENT
ALVERNO COLLEGE
MILWAUKEE, WI 53234-3922

ADVANCED GEOMETRIC CONSTRUCTIONS

Alfred S. Posamentier • **William Wernick**

Printed in the United States of America.
Published simultaneously in Canada.

4 5 6 7 8 9 10 - ML - 99 98 97

Order number DS01803
ISBN 0-86651-429-5

DALE
SEYMOUR
PUBLICATIONS
P.O. BOX 10888
PALO ALTO, CA 94303

INTRODUCTION

All too often geometric constructions are seen as just a collection of techniques using straightedge and compass. In fact, geometric constructions provide one of the richest areas of problem solving in geometry. The solutions of such problems are primarily geometric, requiring few or no skills from other branches of mathematics. Advanced Geometric Constructions presents geometric constructions from a problem-solving perspective. However, each topic is treated comprehensively to provide insight into several key areas of geometric construction.

Chapter I offers a refresher in common construction techniques. These techniques provide the basic tools for the rest of the book. Once the techniques are learned, the trick is to know when to use them. The remaining chapters provide some novel settings for using these construction techniques.

Chapter II applies the basic geometric constructions in unusual and challenging problems, some leading to the development of important mathematical principles.

The triangle constructions presented in Chapter III are derived from a comprehensive list of 179 proposed constructions. These range in complexity from very simple to requiring a fair amount of ingenuity. Solutions are offered to some of these constructions as models for those left for the reader to solve. This chapter provides a particularly rich forum for geometric problem solving.

Chapter IV focuses on the construction of a circle to fit certain restrictions. The treatment of this topic leads to an investigation of "The Problem of Apollonius."

Geometric constructions with special tools are presented in Chapter V. These constructions include: (1) using the compass alone (Mascheroni Constructions), (2) using a straightedge and a fixed circle, (3) using a straightedge and a fixed (immovable) compass, and (4) using a pair of compasses that do not hold the measure when lifted off the paper. Clear, simple language is used to discuss these concepts.

Chapter VI analyzes a construction technique that uses a different type of instrument, a "double straightedge"—that is, a straightedge, such as a ruler, that has two parallel edges.

Exercises are interspersed throughout each chapter to reinforce new concepts as they are presented. The end-of-chapter exercises review the highlights of each chapter.

This book presents geometric constructions in step-by-step detail. As the constructions become more complex, they provide an exciting and challenging arena for solving geometric problems. Many problems are left for the reader to solve, encouraging a study of advanced geometric constructions.

Alfred S. Posamentier
William Wernick
1987

Chapter I

THE BASIC CONSTRUCTIONS

Unless we are told otherwise, we shall do our constructions using only the tools prescribed by the Greek mathematicians of antiquity. These tools are an unmarked straightedge and compasses. The straightedge may be of any length and the compasses may be as large as we find convenient. The Greeks thought of geometry in terms of these two instruments. Their investigation into the types of figures that can be constructed using only these two instruments was quite extensive. In this chapter, we shall study the basic constructions involving the unmarked straightedge and compasses. It will be necessary to master these basic constructions before studying the succeeding chapters.

For convenience, let us agree on a short notation for describing a circle, or an arc of a circle. When we wish to refer to a circle with a center A and radius of length AB, we shall use the symbol (A, AB). The first member of the ordered pair indicates the center of the circle (or arc of the circle) while the second member indicates the length of the radius. The dashed lines in the diagrams are intended to facilitate the proof of the construction and are not part of the actual construction.

Construction 1:

Construct a line segment congruent to another segment.

Given: $\overline{AB}$

Construct: $\overline{CD} \cong \overline{AB}$

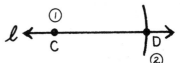

1. Choose any point C on line ℓ
2. Draw arc (C, AB) to meet line ℓ at D
3. $\overline{AB} \cong \overline{CD}$

Construction 2:

a. **Construct a line segment whose length equals the sum of the lengths of two segments.**

Given: $\overline{AB}$ and $\overline{CD}$

Construct: $\overline{EF}$ with

length AB + CD

1. Choose any point E on line ℓ
2. Let arc (E, AB) intersect line ℓ at P.
3. Let arc (P, CD) intersect $\overrightarrow{EP}$ at F.

Thus, EF = AB + CD.

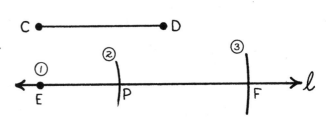

2

b. **Construct a line segment whose length equals the difference of the lengths of two given segments.**

Given: $\overline{AB}$ and $\overline{CD}$

(where AB < CD)

Construct: $\overline{EF}$ with

length CD − AB

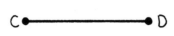

1. Choose any point E on line ℓ
2. Let arc (E, CD) intersect line ℓ at P.
3. Let arc (P, AB) intersect $\overline{EP}$ at F.

Thus, EF = CD − AB.

Construction 3:

Construct an angle congruent to a given angle.

Given: ∠ ABC

Construct: ∠ DEF ≅ ∠ ABC

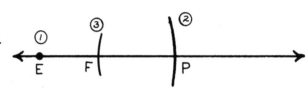

1. Choose any point E on line ℓ
2. With any radius m, let arc (B, m) intersect $\overrightarrow{AB}$ at P, and $\overrightarrow{BC}$ at R.
3. Draw arc (E, m) to intersect line ℓ at D.
4. Let arc (D, PR) intersect arc (E, m) at F.
5. Draw $\overrightarrow{EF}$. Hence, ∠ DEF ≅ ∠ ABC.

Proof Outline: Since $\overline{BR} \cong \overline{ED}, \overline{BP} \cong \overline{EF}$

and $\overline{PR} \cong \overline{FD}$, then

△ PBR ≅ △ FED (s. s. s.).

Therefore, ∠ DEF ≅ ∠ ABC.

Query: How can Construction 3 be used to construct a triangle similar to a given triangle? Are there any other methods available?

Construction 4:

a. **Construct an angle whose measure equals the sum of the measures of two given angles.**

Given: ∠ ABC and ∠ DEF

Construct: ∠ KMN so that

m∠ KMN = m ∠ ABC + m ∠ DEF

1. Choose any point M on line ℓ

2. Construct an angle congruent to $\angle$ ABC with vertex M and a side (ray) contained in line ℓ (see Construction 3), so that $\angle$ NMX $\cong$ $\angle$ ABC.

   (Note: N is on line ℓ.)

3. Construct an angle congruent to $\angle$ DEF, with vertex M and a ray contained in $\overrightarrow{MX}$ so that $\angle$ KMX $\cong$ $\angle$ DEF.

Hence, m $\angle$ KMN = m $\angle$ ABC + m $\angle$ DEF.

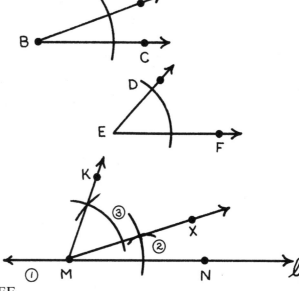

Exercises:

1. How can we use Construction 4a to construct an angle n times the measure of a given angle, where n = 2, 3, 4, 5, . . . ?

2. Construct an angle whose measure is three times m $\angle$ ABC.

b. **Construct an angle whose measure equals the difference of the measures of two given angles.**

Given: $\angle$ ABC and $\angle$ DEF

$\qquad$ (m $\angle$ ABC $<$ m $\angle$ DEF)

Construct: $\angle$ KMN so that

$\qquad$ m $\angle$ KMN = m $\angle$ DEF – m $\angle$ ABC

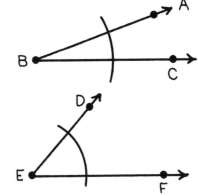

1. Choose any point M on line ℓ

2. Construct an angle congruent to $\angle$ ABC with vertex M and a side (ray) contained in line ℓ (see Construction 3), so that $\angle$ NMX $\cong$ $\angle$ ABC.

   (Note: X is on line ℓ.)

3. Construct an angle congruent to $\angle$ DEF with vertex M and a side (ray) contained in $\overrightarrow{MX}$, so that $\angle$ KMX $\cong$ $\angle$ DEF.

Hence, m $\angle$ KMN = m $\angle$ DEF – m $\angle$ ABC.

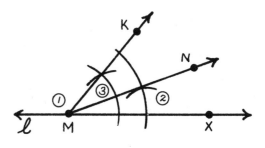

4

Construction 5:

Construct the perpedicular bisector of a given line segment.

Given: $\overline{AB}$

Construct: $\overleftrightarrow{CD}$ so that

$\quad\quad\quad$ $\overleftrightarrow{CD} \perp \overline{AB}$ and

$\quad\quad\quad$ $\overleftrightarrow{CD}$ bisects $\overline{AB}$

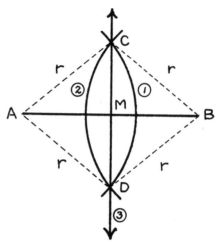

1.   Draw arc (A, r), where $r > \frac{1}{2} AB$

2.   Draw arc (B, r) to intersect arc (A, r)

$\quad\quad$ at points C and D.

3.   Draw $\overleftrightarrow{CD}$.

Thus, $\overleftrightarrow{CD} \perp \overline{AB}$ and $\overleftrightarrow{CD}$ bisects $\overline{AB}$ at M.

Proof Outline: Since AC = BC and AD = BD, C and D are equidistant from the endpoints

$\quad\quad\quad\quad$ of $\overline{AB}$ and hence determine the perpendicular bisector of $\overline{AB}$.

Exercises: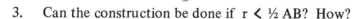

3.   Can the construction be done if $r < \frac{1}{2} AB$? How?

4.   How can you use construction 5 to help construct the median of a given triangle?

5.   Carry through the construction with $r = AB$, and discuss the figure.

Construction 6:

Construct the bisector of a given angle.

Given: $\angle ABC$

Construct: $\overrightarrow{BP}$, the bisector of $\angle ABC$.

1.   Draw arc (B, r) which intersects $\overrightarrow{BA}$ at N and $\overrightarrow{BC}$ at M.

2.   Let arc (M, s) intersect arc (N, s) at P.

3.   Draw $\overrightarrow{BP}$.

Then, $\overrightarrow{BP}$ bisects $\angle ABC$.

Proof Outline: Since NP = MP and BN = BM, $\triangle$ NPB $\cong$ $\triangle$ MPB (s. s. s.).

$\quad\quad\quad\quad$ Therefore, $\angle$ NBP $\cong$ $\angle$ MBP and $\overrightarrow{BP}$ bisects $\angle$ ABC.

Exercises:

6.   Construct the three angle bisectors of a triangle.

7.   In step 2 of Construction 6, could we use *any* radius s? Try radius s, where $s = r$, and

$\quad\quad$ discuss the figure. Try radius t, where $t < r$, and discuss the figure.

8.   Try this construction when $\angle ABC$ is a straight angle, and discuss the figure. Are there

$\quad\quad$ any other related constructions?

Construction 7:

Construct a line perpendicular to a given line and containing a given point not on the given line.

Given: $\overleftrightarrow{AB}$ and point P not on $\overleftrightarrow{AB}$.

Construct: $\overleftrightarrow{PQ} \perp \overleftrightarrow{AB}$.

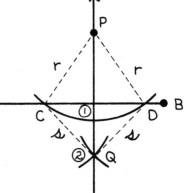

1. Draw arc (P, r) so that it intersects $\overleftrightarrow{AB}$ at C and D.

2. Draw arc (C, s) and arc (D, s) to intersect at Q, where s $>$ ½ CD. We then have $\overleftrightarrow{PQ} \perp \overleftrightarrow{AB}$.

Proof Outline: Since PD = PC and QD = QC, points P and Q are equidistant from the endpoints of $\overline{CD}$ and therefore $\overleftrightarrow{PQ} \perp \overline{CD}$; hence, $\overleftrightarrow{PQ} \perp \overleftrightarrow{AB}$.

Exercises:

9. How does Construction 7 compare with Construction 5?

10. How may Construction 7 be used to construct an altitude of a given triangle?

11. Draw any obtuse triangle, then construct the three altitudes of that triangle. Must we use Construction 7 three times? How can we construct these three altitudes, using Construction 7 only twice?

12. In step 1 of Construction 7, could we use *any* radius r?

13. In step 2 of Construction 7, could we use any radius s, where s $>$ ½ CD? Could s = r?

14. In step 3 of Construction 7, we used Q as one point of intersection of arcs (C, s) and (D, s). Since these arcs must intersect in another point, say T, could we use T instead of Q in this construction?

Construction 8:

Construct a line perpendicular to a given line and containing a given point in the given line.

Given: Point P contained in $\overleftrightarrow{AB}$.

Construct: $\overleftrightarrow{PQ} \perp \overleftrightarrow{AB}$

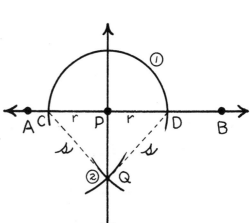

1. Draw arc (P, r) to intersect $\overleftrightarrow{AB}$ at points C and D.

2. Draw arc (C, s) and arc (D, s) to intersect at Q, where s $>$ ½ CD.

3. Draw $\overleftrightarrow{PQ}$. Thus, $\overleftrightarrow{PQ} \perp \overleftrightarrow{AB}$.

Proof Outline: See Proof Outline for Construction 7.

Exercises:

15. Prove that this construction does what it purports to do.

16. How does Construction 8 compare with Constructions 5 and 7?

17. How is Construction 8 related to Construction 6?

18. Could arc (C, s) and arc (D, s) meet above $\overleftrightarrow{AB}$?

Construction 9:

Construct an equilateral triangle with a side of given length.

Given: AB

Construct: Equilateral △ ABC

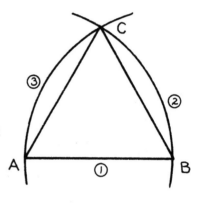

1. Draw $\overline{AB}$.

2. Draw arc (A, AB).

3. Draw arc (B, AB).

4. Let C be the point of intersection of the arcs drawn in steps 2 and 3.

5. Draw $\overline{AC}$ and $\overline{BC}$.

6. △ ABC is equilateral.

Proof Outline: Since AB = AC and AB = BC, △ ABC is equilateral.

Exercises:

19. How can we construct an angle of degree measure 60?

20. How can we construct an angle of degree measure 30?

21. The arcs drawn in step 4 of Construction 9 must meet again at another point, say D. Find D and bring it into the discussion.

Construction 10:

Construct a line parallel to a given line through a given point not on the given line.

Given: $\overleftrightarrow{AB}$, and D not on $\overleftrightarrow{AB}$.

Construct: $\overleftrightarrow{DC} \parallel \overleftrightarrow{AB}$.

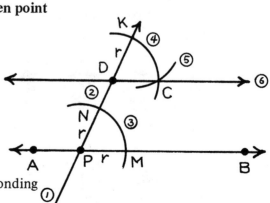

1. Draw any line through D intersecting $\overleftrightarrow{AB}$ at P.

2. Use Construction 3 to make $\angle KDC \cong \angle DPB$. We then have $\overleftrightarrow{DC} \parallel \overleftrightarrow{AB}$.

Proof Outline: $\overleftrightarrow{DC} \parallel \overleftrightarrow{AB}$, since a pair of corresponding angles are congruent.

Exercise:

22. Can you find another method for constructing parallel lines?

Construction 11:

Divide (partition) a line segment into 3 congruent segments.

Given: $\overline{AB}$

Construct: Points M and N on $\overline{AB}$ so that
AM = MN = NB. A similar con-
struction may be used for 2, 3, 4, 5, 6, 7, . . .
congruent segments.

1. Draw any line ℓ containing A.

2. With a convenient radius r, draw arc (A, r)
    meeting line ℓ at C.

3. Draw arc (C, r) meeting line ℓ at D, beyond C.

4. Draw arc (D, r) meeting line ℓ at E, beyond D.

5. Draw $\overline{BE}$.

6. Use Construction 10 to construct lines m and k containing points C and D respectively,
    and parallel to $\overleftrightarrow{BE}$.

7. Let line m intersect $\overline{AB}$ at M, and let line k intersect $\overline{AB}$ at N. We now have

    AM = MN = NB.

Proof Outline: Since AC = CD = DE and m $//$ k $//$ $\overleftrightarrow{BE}$, therefore AM = MN = NB.
                   For if three or more parallel lines intercept segments of equal length on one
                   transversal, then they intercept segments of equal length on any transversal.

Exercises:

23. Divide a line segment into 5 congruent segments.

24. Describe two methods which we can use to divide a line segment into 4 parts of equal
     length.

Construction 12:

**Divide (partition) a given line segment into segments whose lengths are proportional to those
of two given segments.**

Given: $\overline{AB}$, $\overline{NQ}$ and $\overline{RS}$.

Construct: Point P on $\overline{AB}$, so that $\dfrac{AP}{PB} = \dfrac{NQ}{RS}$.

8

1. Draw any convenient line ℓ through A.

2. Draw arc (A, NQ) to intersect line ℓ at C.

3. Draw arc (C, RS) to intersect ℓ at D, beyond C.

4. Draw $\overline{BD}$.

5. Use Construction 10 to construct a line, m, through C and parallel to $\overline{BD}$.

6. The intersection of line m and $\overline{AB}$ is point P. We then have $\dfrac{AP}{PB} = \dfrac{NQ}{RS}$.

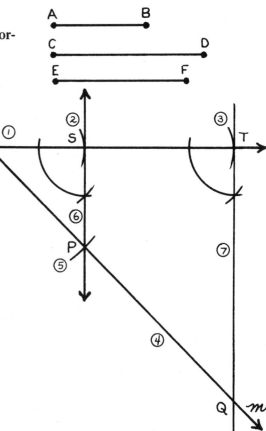

Proof Outline: If a line is parallel to one side of a triangle, it divides the other two sides proportionally. Therefore, $\dfrac{AP}{PB} = \dfrac{AC}{CD}$. But AC = NQ and CD = RS; hence, $\dfrac{AP}{PB} = \dfrac{NQ}{RS}$.

Exercise:

25. How might this construction be done if NQ = RS? If NQ = 2RS?

Construction 13:

Construct the line segment whose length is the fourth proportional to the length of three given line segments.

Given: $\overline{AB}, \overline{CD}$ and $\overline{EF}$.

Construct: $\overline{PQ}$, so that $\dfrac{AB}{CD} = \dfrac{EF}{PQ}$.

1. Choose any convenient point R on line ℓ

2. Draw arc (R, AB) to intersect line ℓ at S.

3. Draw arc (S, CD) to intersect line ℓ at T ,beyond S.

4. Draw any convenient line m through R.

5. Draw arc (R, EF) to intersect line m at P.

6. Draw $\overleftrightarrow{SP}$.

7. Through point T, construct line k $/\!/$ $\overleftrightarrow{SP}$. (See Construction 10.)

8. Line k meets line m at point Q. Since $\dfrac{RS}{ST} = \dfrac{RP}{PQ}$, $\overline{PQ}$ is the required segment.

Proof Outline: See Proof Outline for Construction 12.

Exercise:

26. Prove that Construction 13 does what it purports to do.

Note: To construct a line segment whose length is the product of the lengths a and b of two given segments, construct the fourth proportional to lengths 1, a and b (1 is an agreed-upon unit length). If x is the length to be constructed, we then have $\frac{1}{a} = \frac{b}{x}$ and x = ab.

Note: To construct a line segment whose length is the quotient of the lengths a and b of two given segments, construct the fourth proportional to lengths b, a and 1 (1 is an agreed-upon unit length). If y is the length to be constructed, we then have $\frac{b}{a} = \frac{1}{y}$ and $y = \frac{a}{b}$.

Exercises:

27. Is it possible to construct a segment of length $\frac{a+b}{a \cdot b}$, if the unit length and the lengths a and b are given? If so, describe how.

28. Construct a segment of length $\frac{ab}{a+b}$.

Construction 14:

Construct the segment whose length is the mean proportional between the lengths of two given segments.

Given: $\overline{AB}$ and $\overline{CD}$.

Construct: $\overline{PQ}$, so that $\frac{AB}{PQ} = \frac{PQ}{CD}$.

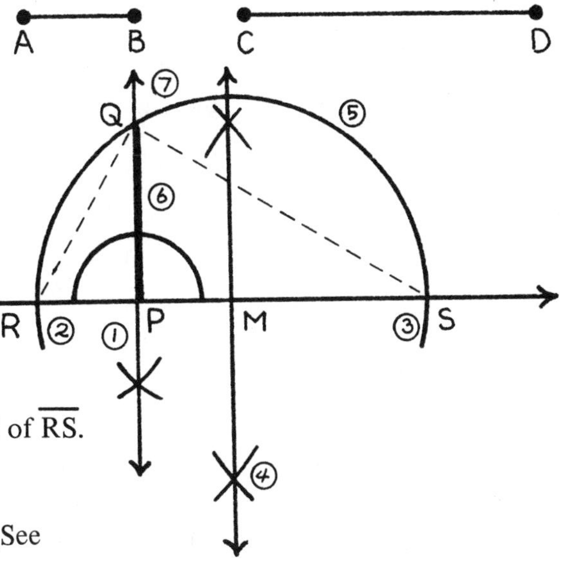

1. Choose any point P on a convenient line ℓ

2. Draw arc (P, AB) to intersect line ℓ at R.

3. Draw arc (P, CD) to intersect $\overrightarrow{RP}$ at S, so that P is between R and S.

4. Construct the perpendicular bisector of $\overline{RS}$ (see Construction 5) to locate the midpoint M of $\overline{RS}$.

5. Draw arc (M, MR).

6. Construct the line perpendicular to $\overleftrightarrow{RS}$ at P. (See Construction 8.)

7. Let Q be the intersection of this line and arc (M, MR). Hence, PQ is the mean proportional between AB and CD.

Proof Outline: $\angle$ RQS is a right angle since it is inscribed in a semicircle. $\overline{PQ}$ is the altitude on the hypotenuse of right $\triangle$ RQS. It follows that PQ is the mean proportional between PR and PS, and therefore between AB and CD.

Exercise:

29. If we are given a segment of unit length, how may we construct a line segment of length $\sqrt{2}$ or $\sqrt{3}$ or $\sqrt{5}$ or $\sqrt{7}$, etc.? Can you discover other methods?

Construction 15:

Locate the center of a given circle.

Given: A circle.

Construct: The center of the circle.

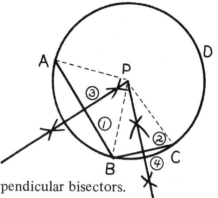

1. Take any three points, A, B, C, on the circle and draw chords $\overline{AB}$, $\overline{BC}$.

2. Construct the perpendicular bisector of chords $\overline{AB}$, $\overline{CB}$ (as in Construction 5).

3. Let P be the point of intersection of these two perpendicular bisectors.

4. P is the center of the circle.

Proof Outline: Since any point in the perpendicular bisector of a line segment is equidistant from the endpoints of that segment, PA = PB = PC. Hence, P is the center of the circle through A, B, C, since it is equidistant from them all.

Exercises:

30. Construction 15 solves what used to be called "The Broken Millstone Problem," in which we were required to construct a whole circle, given just a piece of its circumference. Discuss this.

31. Discuss the related construction if we start with four points, A, B, C, D, and use the chords $\overline{AB}$ and $\overline{CD}$.

32. The circle and the line are the *only* two plane curves of constant curvature, which means that any piece will fit along any other piece. Thus, a line segment can be "extended to any length." Show how any arc of a circle can be used alone to draw the whole circumference, without finding the center of the circle.

Construction 16:

Construct the circumscribed circle of a given triangle.

Given: △ ABC

Construct: The circumscribed circle of △ ABC.

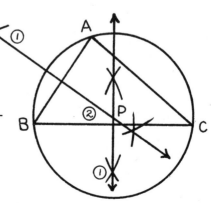

1. Construct the perpendicular bisector of $\overline{AB}$ and $\overline{BC}$ (see Construction 5).

2. Let P be the point of intersection of these perpendicular bisectors.

3. Draw circle (P, PB), the required circle.

Proof Outline: See Proof Outline for Construction 15.

Exercises:

33. Prove that Construction 16 does what it purports to do.

34. How does Construction 16 compare with Construction 15?

35. Why can we be sure that the perpendicular bisectors in Construction 16 actually meet?

Construction 17:

Construct the inscribed circle of a given triangle.

Given: △ ABC

Construct: The inscribed circle of △ABC.

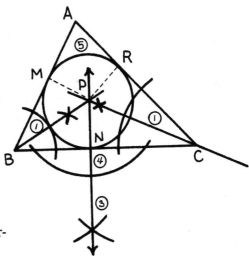

1.   Construct the bisectors of ∠ B and ∠ C.

2.   Let P be the point of intersection of these angle bisectors.

3.   Construct the line which contains point P and is perpendicular to $\overline{BC}$. (See Construction 7.)

4.   Let N be the point of intersection of this perpendicular line and $\overline{BC}$.

5.   Draw circle (P, PN), the required circle.

Proof Outline: Since every point on the bisector of an angle is equidistant from the sides (rays) of the angle, P is equidistant from $\overleftrightarrow{AB}$ and $\overleftrightarrow{BC}$ as well as from $\overleftrightarrow{AC}$ and $\overleftrightarrow{BC}$. N will be the point of tangency of $\overleftrightarrow{BC}$ and the required circle, since a radius is perpendicular to the tangent $\overleftrightarrow{BC}$ at the point of contact. If M and R are the other points of tangency of the required circle with the sides of the triangle, then PN = PM = PR. Therefore, the circle (P, PN) is the inscribed circle of △ ABC.

Exercises:

36. Can this construction be used to find the inscribed circle of an obtuse triangle?

37. If we extend the sides of any triangle, can we use a construction like Construction 17 to construct a circle tangent to one side of the triangle and the extensions of the other two sides?

38. How many circles can be constructed tangent to any two lines?

39. How many circles can be constructed tangent to any three lines?

40. Discuss cases other than those in exercises 38 and 39.

Construction 18:

Construct a line tangent to a given circle at a given point on the circle.

Given: Circle (Q, PQ).

Construct: $\overleftrightarrow{PR}$ tangent to Circle (Q, PQ).

1.   Draw $\overleftrightarrow{QP}$.

2.   Construct a line which contains P and is perpendicular to $\overleftrightarrow{QP}$. (See Construction 8.)

3. $\overleftrightarrow{PR}$ is tangent to circle (Q, PQ).

Proof Outline: Since $\overleftrightarrow{PR}$ is perpendicular to radius $\overline{PQ}$ at P, it is tangent to the Circle (Q, PQ).

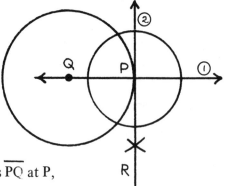

Construction 19:

Construct two tangents to a given circle and containing a point in the exterior of the circle.

Given: Point P in the exterior of circle (Q, RQ).

Construct: $\overleftrightarrow{PA}$ and $\overleftrightarrow{PB}$ tangent to circle (Q, QR).

1.   Draw $\overleftrightarrow{QP}$.

2.   Construct the perpendicular bisector of $\overline{PQ}$ to find the midpoint M of $\overline{PQ}$. (See Construction 5.)

3.   Draw circle (M, MQ). A and B are the points where this circle intersects the given circle.

4.   Draw $\overleftrightarrow{PA}$ and $\overleftrightarrow{PB}$, the required lines.

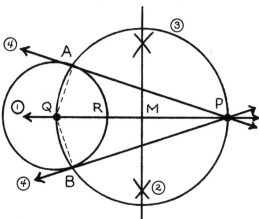

Proof Outline: $\angle$ PAQ and $\angle$ PBQ are right angles since they are each inscribed in a semicircle. $\overleftrightarrow{PA}$ and $\overleftrightarrow{PB}$ are tangents, since they are each perpendicular to a radius of the circle at a point on the circle.

Note: This construction allows us to draw a tangent line only when we know two points on it. We started with point P, then found point A, then drew a tangent line. *Do not* confuse this accurate construction with the usual easy sketch, which *looks* accurate; that is, to place the straightedge so that point P appears at its edge, then move the straightedge until you can just about see a tiny mark along the edge, then draw a "tangent" line through P and this tiny mark of the circle.

Constructibility

So far in this chapter we have performed many basic geometric constructions. Yet there are three construction problems of great historical interest which have resisted solution over the centuries. These three problems which follow are known as the "Three Problems of Antiquity."

I. Trisect a given arbitrary angle;

II. Construct the edge of a cube whose volume is twice that of a given cube;

III Construct a square whose area equals that of a given circle.

For over two thousand years mathematicians attempted to solve these problems. Although these attempts at the solution failed, many interesting concepts of mathematics were discovered as by-products. It took the accumulated genius of these mathematicians to enable us, within the last two centuries, to prove that these problems can in fact not be solved with just a straightedge and compasses. These proofs have their foundations in Algebra and Analysis.

To present the proofs here in a rigorous fashion would not only be too lengthy for this small book but also beyond its scope. We cheerfully direct the ambitious reader to the following books, where the proofs can be found.

1. Felix Klein, *Famous Problems of Elementary Geometry*. Chelsea Publishing Company, New York, 1962. (This book contains a very comprehensive proof of the impossibility of the solution of each of the three problems under the given conditions.)

2. Nicholas D. Kazarinoff, *The Ruler and the Round*. Prindle, Weber & Schmidt, Boston, 1970. (This book presents complete proof of the impossibility of the solution of the first two problems above. The book was chosen for this list because its comprehensive treatment is written in a pleasant and easily understood style.)

3. Richard Courant and Herbert Robbins, *What Is Mathematics?* Oxford University Press, New York, 1941, pp. 127-140. (This book treats the first two problems in a very concise manner, but only outlines the procedure needed to prove the impossibility of the solution of the third problem.)

4. Edwin E. Moise, *Elementary Geometry from an Advanced Standpoint*. Addison-Wesley, Reading, Massachusetts, 1974, pp. 227-241. (This book offers an interesting presentation of the proofs of the impossibility of the solution of the first two problems, using modern terminology and notation.)

These are a few of the many books available which present proofs of the impossibility of the solutions of the three problems and are written in a manner intelligible to a student in the tenth grade.

We must remember that these solutions are impossible when their solution is attempted using only an unmarked straightedge and a pair of compasses. As soon as we make a slight alteration of these tools, their solution may become possible. For example, if we allow the use of just two markings on the previously unmarked straightedge, the problem of trisecting an arbitrary angle becomes solvable. The following scheme for trisecting an arbitrary angle using a pair of compasses and straightedge with just two markings on it was first presented by Archimedes.

Let $\angle$ ABC be an arbitrary angle.

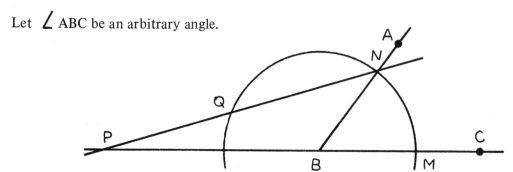

Draw arc (B, BM), where BM is a convenient length. Make two markings, P and Q, on your straightedge so that PQ = BM. Then place the marked straightedge so that it passes through N, the intersection of $\overrightarrow{BA}$ and arc (B, BM), and so that Q is on arc (B, BM) and P is on $\overleftrightarrow{BC}$. The angle so formed, $\angle$ NPM, has a measure of one-third that of $\angle$ ABC.

Exercises:

41.   How would we then use this construction to trisect $\angle$ ABC?

42.   Prove that m $\angle$ NPM = 1/3 m $\angle$ ABC.

There are many other schemes for trisecting an angle, using tools other than just an unmarked straightedge and compasses. There are also procedures for solving the other two problems; however, these procedures also utilize tools other than the unmarked straightedge and compasses.

For the remainder of this book, we shall concern ourselves only with those constructions which are possible using an unmarked straightedge and compasses. You will find that these rather limited tools have many fascinating applications.

Chapter II

APPLICATIONS OF BASIC CONSTRUCTIONS

Now that we have the basic constructions of Chapter I, we can start putting them together in more complicated ways to solve more difficult and more interesting problems.

I. Geometric Solutions of "Algebraic" Equations.

1. The equation $ax + b = 0$ has a simple algebraic solution: $x = \frac{-b}{a}$. If a and b are lengths, how shall we interpret the above equation and solution? A comment on "dimension" is necessary here. If we are to interpret a, b and x as lengths, we assign to them each the dimension 1, since these lengths can be assigned to segments on a number line. As soon as we multiply two lengths, however, we must deal with a quantity of dimension 2, for example, an area. (Remember that the area of a rectangle is ab, and the area of square is s^2.) Since there is not much meaning attached to the "sum" of an area and a length, we agree, from now on, that when these equations are to be given geometric interpretations, we may add only lengths to lengths, or areas to areas, and so on. That is, all terms in any such "geometric equation" must be of the same degree, i.e., such an equation must be homogeneous.

In the equation, $ax + b = 0$, we have a mixture: ax is of the second degree and b is of the first degree; thus, this equation has no geometric interpretation. The solution, $x = \frac{-b}{a}$, is of zero degree and has no geometric representation.

We could make a small change, however, if we introduce a unit length 1 and rewrite the equation thus: $ax + 1b = 0$, for which the solution is: $x = \frac{-1b}{a}$. Now we have a routine geometric exercise: we interpret lengths in their absolute sense, representing both 1 and –1 by the same unit segment, then simply construct x as the fourth proportional (see Construction 13 in Chapter I) to the lengths a, b and 1, since the proportion $\frac{a}{b} = \frac{1}{x}$ leads to the same "solution" for x.

2. Solve geometrically: $x = 3a$.

In this context, we interpret "3a" to mean $a + a + a$, which is a sum of three lengths and therefore of dimension 1. We leave to you the simple construction for finding a segment three times as long as a given segment.

In this category, we could construct $\frac{1}{3}$ a, $\frac{5}{7}$ a, and so on. But be careful of x = ba, because, if b and a are both lengths, then this term is of dimension 2 and x cannot be directly represented as a length.

3. Solve geometrically: x = ba.

Since the product is of dimension 2, we must introduce a unit length in order to represent x as a quantity of dimension 1, thus: $x = \frac{ba}{1}$. Then we can easily write: $\frac{1}{a} = \frac{b}{x}$ and construct x as the fourth proportional to 1, a and b.

4. Solve geometrically: $x = \frac{abc}{de}$.

The right side of this equation is of degree 1, i.e., (3 − 2), so we should be able to construct a segment of length x. We do it in two stages: write $x = \left(\frac{ab}{d}\right)\left(\frac{c}{e}\right)$, and write $y = \frac{ab}{d}$, therefore: $x = \frac{yc}{e}$. You should see now that we can construct y from the proportion $\frac{d}{b} = \frac{a}{y}$; and once we have y, we can then construct x from the proportion $\frac{e}{c} = \frac{y}{x}$.

Exercises

Solve each of the following geometrically, with given lengths a, b, c, d, e, and any convenient unit length 1:

1. Carry through Construction 4.

2. $x = \frac{a^2}{b}$

3. $x = \frac{a^2 b^2}{cde}$

4. $x = a^2$

5. $x = a^3$

   (Hint: $a^3 = \frac{a^3}{1^2}$.)

6. $x = \frac{2a + 3b}{4c}$

7. $x = \frac{(a + b)c}{b}$

8. $x = \frac{1}{a}$ (Hint: $\frac{1}{a} = \frac{1 \cdot 1}{a}$ Another solution will be given in Chapter III.)

9. $x = \frac{a}{bc}$

10. $x = \frac{cd}{c^2 d}$

5. Solve geometrically: $x^2 - ab = 0$.

Algebraically, we have $x = \sqrt{ab}$, and we recognize x as the mean proportional between a and b, as found in Chapter I (Construction 14). We see again that, since ab is of dimension 2 and its square root is thus of dimension 1, we can construct a segment whose length is $\sqrt{ab}$.

6. Solve geometrically: $x^2 - a = 0$.

Since we have $x = \sqrt{a}$, we must introduce a unit length thus: $x = \sqrt{a \cdot 1}$ and then find x as the mean proportional between the unit segment and the segment of given length a.

7. Solve geometrically: $x^2 = a^2 + b^2$. This is our old friend, the Pythagorean Formula.

On the sides of any right angle, lay off segments CA = b and CB = a. Then, of course, the hypotenuse AB = x = $\sqrt{a^2 + b^2}$.

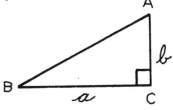

If we had been given $x^2 = c^2 - d^2$, we would have found the segment of length x as one leg of the right triangle whose hypotenuse and other leg have lengths c and d, respectively.

8. Solve geometrically: $x^2 = a + bc$.

We adjust the dimensions with a unit length, as usual, writing $x^2 = 1 \cdot a + bc$; then we find solutions to $y^2 = 1 \cdot a$ and $z^2 = bc$ by mean proportionals, and finally find $x^2 = y^2 + z^2$ from a right triangle as we have just done.

Exercises

Solve each of the following geometrically:

11. $x = \sqrt{abc}$ (Hint: $abc = \frac{abc}{1}$)

12. $x = a + bc$

13. $x = a^2 + bc$

14. $x = a + \dfrac{1}{a}$

15. $ax + b = cx + d$

16. $ax^2 = bc$

17. $a^2x = bc$

18. $a^2x^2 = bc$

19. $x = \dfrac{1}{\sqrt{a}}$

20. $x = \dfrac{1}{a} + \dfrac{1}{\sqrt{a}}$

9. Solve geometrically: $x^2 + bx = c^2$.

We are moving up to the general quadratic equation in x. The given equation is equivalent to $x(x + b) = c^2$, and should recall a theorem about a tangent and secant to the same circle from the same external point. (What theorem is that?)

This solution is completely done in Chapter III, where it is essential in the solution to problem 105, so we will not present it here.

Exercises

21. Try to work out the above solution by yourself, before you look it up.

Solve each of the following geometrically:

22. $x^2 + x = c^2$ (Hint: $x = 1x$.)

23. $x^2 + bx = c$

24. $x^2 + bx + c = 0$ (Hint: Use the same representation for $c = 1c = -1c$.)

25. $ax^2 + bx + c = 0$ (Hint: Adjust all dimensions up to 3, thus: $bx = 1 \cdot bx$ and $c = 1^2c$.)

18

II. Locus Relations.

1. Find all the points which are equidistant from two fixed points A and B.

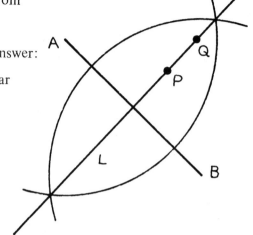

This familiar problem has a familiar answer: All these points will be on the perpendicular bisector of $\overline{AB}$, as shown. We analyze this situation in greater detail now: Two theorems are essentially involved.

(a) If P is any point on the perpendicular bisector of $\overline{AB}$, then P is equidistant from A and B.

(b) If Q is any point which is equidistant from A and B, then Q is on the perpendicular bisector of $\overline{AB}$. In this case, both of these theorems are true. (Can you prove this?) They are converses of each other and together establish the fact that "The *locus* of points equidistant from two fixed points A and B is the perpendicular bisector of $\overline{AB}$."

Locus relations play an important role in the rest of this book and in much of geometry, so we develop those ideas in more detail now.

Two essential ingredients always appear in locus discussions: (1) A *condition* with respect to some given information; (2) a *locus* which is a set of points (a line, a circle, a region or parts or combinations of these). These two ingredients *must* be linked by this pair of relations:

a. *Every* point that satisfies the *condition* must lie on the *locus*; and conversely:

b. *Every* point that lies on the *locus* must satisfy the *condition*.

In our first illustration above, the *condition* was that of being equidistant from A and B, and the *locus* was the perpendicular bisector of $\overline{AB}$. Since *both* theorems were true (though we did not prove them): condition $\longrightarrow$ locus and locus $\longrightarrow$ condition, we have a valid locus relation here.

2. Circumscribe a circle about a given $\triangle$ ABC.

Since the center O of the required circle must be equidistant from all three vertices, it must, in particular, be equidistant from A and B. That is, one locus for O is L_1, the perpendicular bisector of $\overline{AB}$. Analogously, another locus for O is L_2, the perpendicular bisector of $\overline{BC}$. Since O must be on *both* L_1 and L_2, it must be on their intersection and thus we locate O. As an extra dividend, which you may already know, since O is equidistant from A and B, and also from B and C, it must therefore be equidistant from A and C — which means that it must lie on the perpendicular bisector of $\overline{AC}$. Thus we have shown that the perpendicular bisectors of the three sides are concurrent at the point O.

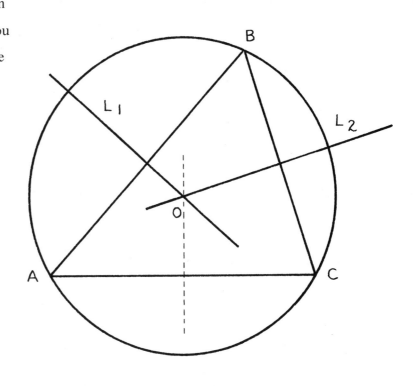

3. What is the locus of points which are 1 unit away from a given line $\overleftrightarrow{AB}$?

Consider the line L_1, parallel to $\overleftrightarrow{AB}$ and one unit above it, as shown. Certainly, every point that lies on L_1 *is* one unit away from $\overleftrightarrow{AB}$, and *part* of the locus requirement is satisfied. But, if we consider only L_1, then not every

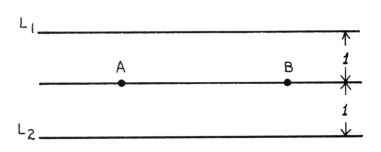

point which is one unit from $\overleftrightarrow{AB}$ lies on L_1; therefore, L_1 alone is *not* the desired locus.

The locus for the given condition is the *pair* of lines, L_1 and L_2, which are both parallel to $\overleftrightarrow{AB}$, and one unit from $\overleftrightarrow{AB}$ on each side of it. Now, with this pair of lines we have *both*:

   (1)   Every point which is one unit from $\overleftrightarrow{AB}$ lies on one of the parallel lines L_1, L_2.

   (2)   Every point on one or the other of the parallel lines is one unit from $\overleftrightarrow{AB}$.

4. What is the locus of points equidistant from two given lines

   a. when the lines are parallel? (We leave this solution to you.)

   b. when the lines intersect? (We state the result and leave the proofs to you.)
      The pair of lines which bisect the angles formed by the given lines. What is
      the relation of these two angle bisectors to each other?

Exercises

26. Show how the results of problem 4 above lead to the inscribed circle of a triangle.

27. Given *any* three lines, show how to construct a circle tangent to all of them.
    (Consider the case with some lines parallel.)

28. Given any three lines which intersect in three distinct points. How many circles
    can be drawn that are tangent to all three lines? (This is discussed in detail in
    Chapter IV.)

29. Given any point P within a
    given angle $\angle$ ABC, con-
    struct segment $\overline{QPR}$, with
    its ends on the sides of the
    angle, and which is bisected
    by point P.

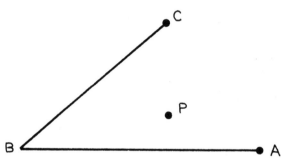

30. Given line ℓ, and point P not on ℓ. Find the locus of points $\{Q_i\}$ such that
    ℓ bisects segment $\overline{PQ_i}$ (where $i = 1, 2, 3, \ldots$).

31. In $\triangle$ ABC, find the locus of the midpoints of all segments parallel to $\overline{BC}$, and
    whose endpoints are on the sides of $\angle$ BAC.

32. Find the locus of the third vertices of all triangles which have the same base $\overline{BC}$
    as given $\triangle$ ABC, and the same area as $\triangle$ ABC.

33. Find the locus of the centers of all circles of radius a which are tangent to a given
    line ℓ.

34. Find the locus of the centers of all circles which are tangent to a given line ℓ at
    a given point P on ℓ.

35. Find the locus of all points, the sum of whose distances from two given perpendicu-
    lar lines, ℓ_1 and ℓ_2, is a given constant a.

One more locus is much used in later construction problems.

5. What is the locus of all points that subtend a given angle on a given segment? We make the problem more specific: Given segment $\overline{AB}$ and $\angle$ CDE, find the locus of all points $\left\{ P_i \right\}$ so that $\angle AP_iB \cong \angle$ CDE. (i = 1, 2, 3, . . .)

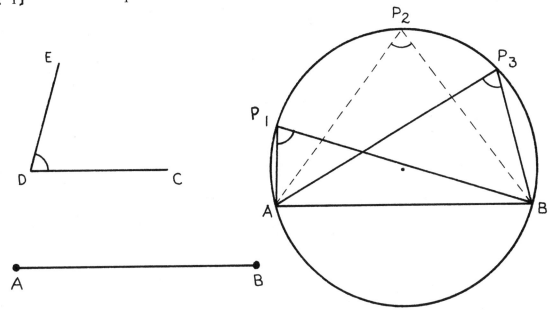

The figure should make the locus clear. It is the circular arc $\overset{\frown}{AP_iB}$. If any one of the angles $\angle AP_iB$ is congruent to $\angle$ CDE, then so are the others, since all three are inscribed angles that intercept $\overset{\frown}{AB}$ and therefore congruent to each other. A construction for this locus would start with segment $\overline{AB}$.

One locus for the center of the desired circle is L_1, the perpendicular bisector of $\overline{AB}$. Now, at the end of the $\overline{AB}$, we construct $\angle$ BAC congruent to the given $\angle$ CDE. If we consider this angle as a tangent-chord angle of the desired circle, then another locus for its center is L_2, the perpendicular to $\overleftrightarrow{AC}$ at A. Then L_1 and L_2 intersect at O, the desired center and finally the desired locus is part of the circle (O, OA).

If P is any point on the upper arc $\overset{\frown}{AB}$, then m $\angle$ APB = ½m $\overset{\frown}{AB}$ = m $\angle$ BAC, as desired. We note that the locus is just the upper arc of this final circle; if Q is any point on the lower $\overset{\frown}{AB}$, then you should have no trouble in showing that m $\angle$ AQB equals the measure of the supplement of $\angle$ BAC which equals the measure of the supplement of given $\angle$ CDE.

An immediate and much used consequence of this locus is the fact that the locus of all points P that subtend a right angle on a given segment $\overline{AB}$ is the circle with diameter $\overline{AB}$, as shown. We leave the proof to you.

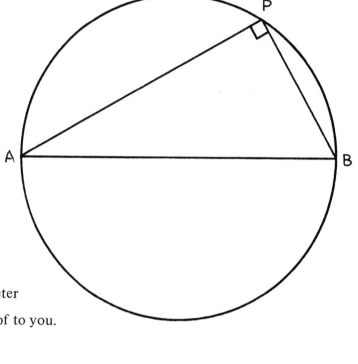

III. Regular Polygons.

This beautiful figure is one of the easiest to draw, and always delights the student when he first discovers it. The hexagon ABCDEF is regular, that is, it has six congruent sides and six congruent angles. The figure is rich also in equilateral triangles. We have drawn one of these by connecting alternate vertices. How many others can you draw easily in the figure?

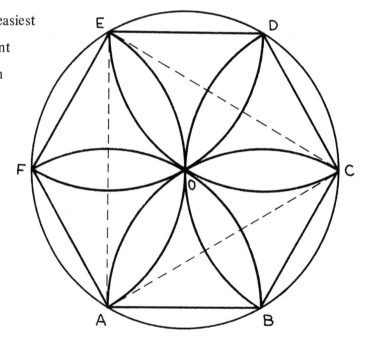

1. Construct a square (a) given a side, s; (b) given a diagonal, d. We leave these two familiar constructions to you, with the hint that a circle would be useful in the second, but not in the first.

2. Construct a regular dodecagon (12 sides). Start with the hexagon and go on by yourself.

3. Construct a regular octagon (8 sides). Start with a square.

Exercises

Construct angles with degree measure:

36.	90	37.	60	38.	30
39.	45	40.	22½	41.	15
42.	7½	43.	67½	44.	105

Consider the measures of various angles in the polygons constructed earlier; don't forget how to add or subtract angles, or how to bisect them.

4. Construct a regular pentagon (5 sides). This is a much more difficult and beautiful problem, for which several solutions are available. The following solution is quite straightforward and starts with an analysis of a regular decagon (10 sides). Then, when that has been constructed, we can easily get the pentagon by joining alternate vertices of the decagon.

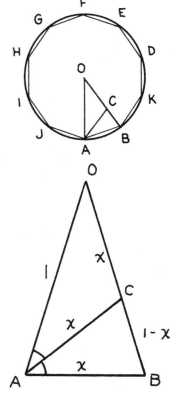

Suppose such a regular decagon has been inscribed in a unit circle (radius = 1) and that we have drawn radii $\overline{OA}$, $\overline{OB}$ to the ends of one side $\overline{AB}$. Then, in this isosceles $\triangle$ AOB we have OA = OB = 1, and m$\angle$ AOB = 36° (Why?).

Then each base angle will have measure 72° and $\overline{AC}$, the bisector of $\angle$ BAO, will form two isosceles triangles, $\triangle$ AOC and $\triangle$ BAC. (Show why.) Thus, BA = AC = CO = x and CB = 1 – x. But, since $\triangle$ AOB $\sim$ $\triangle$ BAC (Why?), we have the proportion $\frac{1}{x} = \frac{x}{1-x}$, which leads to the equation $x^2 + x - 1 = 0$. This equation has two roots, one of which has geometric significance: x = ½($\sqrt{5}$ – 1), since it is the length of a segment that we can construct, as follows (see next page):

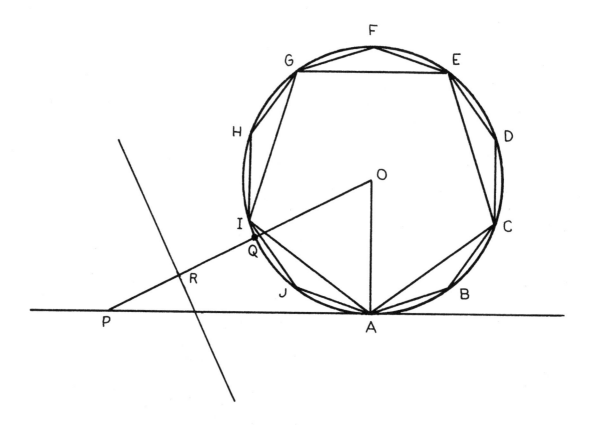

At any point A of a line, erect a perpendicular of length $1 = OA$, and construct the unit circle $(O, 1)$ tangent to that line at A. On the line, make $AP = 2$ and then draw $\overline{OP}$. Then $OP = \sqrt{5}$ and $PQ = \sqrt{5} - 1$. Finally, the perpendicular bisector of $\overline{PQ}$ gives us $QR = \frac{1}{2}(\sqrt{5} - 1) = x$, and we can now lay off on the circle: $x = QR = AB = BC = CD = DE = \ldots$. Of course, the regular pentagon is found by joining alternate vertices A, C, E, G, I.

We have now indicated the construction of regular polygons with 3, 4, 5, 6, 8 and 10 sides. You should have no difficulty (at least theoretically) with 12, 24, 48, . . . sides, or with 10, 20, 40, . . . sides (why?). What can we do about a regular heptagon (7 sides)?

The great Greek mathematicians, and many great mathematicians after them, tried to construct regular polygons of 7, 9, 11, 13, . . . sides with the usual Euclidean tools — compasses and straightedge — but could not do so. Then in 1796, a young man of 18 solved these problems once and for all — not by doing them, but by showing that they were impossible to do. What Carl Friedrich Gauss did on March 30, 1796 was to prove for the first time the basic theorem on the constructibility of regular polygons with compasses and straightedge. There is no difficulty with $n = 2^k$, when $k > 1$, where n is the number of sides of the regular polygon, because we already know how to construct regular polygons of 4, 8, 16, . . . sides. When n is any even number, 2m, then the n-sided polygon can be obtained by bisecting the arcs on the sides of the m-sided circumscribed polygon, so the whole question boils down to what happens when n is *odd.*

This is the question that Gauss answered finally; with the following beautiful theorem:

A regular polygon with an odd number of sides, n, can be constructed with compasses and straightedge if and only if n is a prime number of the form $2^{2^x} + 1$, or a product of *different* prime numbers of that same form. (A prime number is an integer that has only itself and 1 as integral divisors, thus, 7, 11 and 13 are primes, but 6, 12 and 14 are not.)

The proof is difficult and will not be given here, but we'll discuss some consequences. If $x = 0$ in $2^{2^x} + 1$, then $n = 2^{2^0} + 1 = 2^1 + 1 = 3$, and we know that we can construct an equilateral triangle. When $x = 1$, then $n = 2^{2^1} + 1 = 5$ and we have just seen that we can construct the regular pentagon. When $x = 2$, then $n = 2^{2^2} + 1 = 17$. Since 17 is a prime number, Gauss had discovered that a regular 17-sided polygon is constructible with compasses and straightedge, something that was completely unknown to earlier mathematicians.

The numbers of the form $2^{2^x} + 1$ are called "Fermat's numbers" after Pierre Fermat, who worked with them in the seventeenth century. The first three of them, as we have seen, are 3, 5, 17. We then have $2^{2^3} + 1 = 257$, which is also a prime, and $2^{2^4} = 65,537$, which is also a prime; but $2^{2^5} + 1 = 4,294,967,297$ is *not* a prime and $2^{2^6} + 1 = $ the enormous number 18,446,744,073,709,551,617, is also *not* a prime. If you are impressed with large numbers, these should impress you. They have little further use. Fermat thought, mistakenly, that these numbers would all be prime, but Gauss had made no error. When $n = 257$ or $n = 65,537$, since these are prime Fermat numbers, it *is* possible to construct regular polygons with those numbers of sides, a result of theoretical, but no practical significance at all. When $n = 9 = 3 \cdot 3$, since n is not a product of *different* Fermat primes, then that polygon is *not* constructible. When $n = 15 = 3 \cdot 5$, since n *is* now a product of different Fermat primes, then the theory says that a 15-sided regular polygon *can* be constructed, and indeed the Greek mathematicians knew this and did this.

Exercise

45. Construct a regular 15-sided polygon. (Hint: Its central angle must be $1/15 (360^\circ) = 24^\circ$ in measure; and $24 = 60 - 36$; and 60° and 36° are measures of central angles for other regular polygons.)

This finishes our discussion of the constructibility of regular polygons, whose theory is completely known in the sense that we can tell positively which can and which cannot be constructed. However, there is still an essential step between knowing that something can be done and knowing how to do it. In our analysis of the decagon and pentagon, we solved a quadratic equation and then constructed a segment whose measure was one of the roots. An analysis of the 17-sided regular polygon would lead to the solution of the 16th degree equation and a lot of complicated algebra which would be out of place here, so we omit it, and omit also any further details on constructibility.

IV. Miscellaneous Problems.

We close this chapter with a section on some useful but rather miscellaneous and attractive problems.

1. Find a triangle that has the same area as a given polygon.

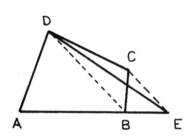 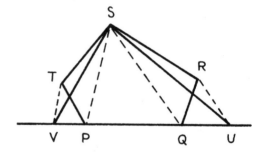

We illustrate with a quadrilateral ABCD, and then a pentagon PQRST, after which the general procedure should be clear. In quadrilateral ABCD, draw diagonal $\overline{BD}$, then a line $\overleftrightarrow{CE}$ parallel to $\overleftrightarrow{BD}$, which intersects $\overleftrightarrow{AB}$ at E. Then, $\triangle$ ADE has the same area as quadrilateral ABCD.

The proof follows from the fact that $\triangle$ BCD and $\triangle$ BED have the same area, since they have the same base $\overline{BD}$ and congruent altitudes, since C and E, by construction, lie on a line parallel to that base. Then, in terms of areas, quadrilateral ABCD = $\triangle$ ABD + $\triangle$ BCD = $\triangle$ ABD + $\triangle$ BED = $\triangle$ AED, as required.

With the pentagon, we do the same sort of thing twice. We make $\overleftrightarrow{RU}$ parallel to diagonal $\overline{QS}$, and $\overleftrightarrow{TV}$ parallel to diagonal $\overline{SP}$, as shown. Then, much as above, by combining areas, we have eventually the area of pentagon PQRST equal to the area of $\triangle$ USV.

If we started with more sides, we simply slide off a corner at a time, as often as necessary until we get down to the triangle we want.

Exercises

46. Construct a triangle that has the same area as a given hexagon.

47. Construct a right triangle that has the same area as a given triangle.

48. Construct an isosceles triangle that has the same area as a given triangle.

49. Construct an isosceles right triangle that has the same area as a given triangle. (Hint: Use algebraic analysis.)

50. Construct a square with the same area as a given triangle.

51. Construct a square with the same area as a given quadrilateral. (Hint: Use problem 50, just above.)

52. Construct a square with the same area as a given pentagon.

53. Construct an equilateral triangle that has the same area as a given triangle. (Hint: Use algebraic analysis.)

54. Construct an equilateral triangle that has the area of a given quadrilateral.

55. Construct an equilateral triangle with an area of 6 square inches.

56. Given △ ABC and △ DEF. Construct △ PQR to be similar to △ ABC and to have the same area as △ DEF.

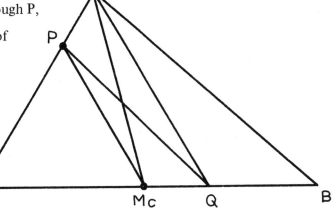

2. Point P lies on a side of △ ABC. Through P, draw a line which will bisect the area of that triangle.

Suppose P lies on $\overline{AC}$. Draw the median $\overline{CM_c}$ and line $\overleftrightarrow{PM_c}$. Then, through C, draw $\overleftrightarrow{CQ}$ parallel to $\overleftrightarrow{PM_c}$, to intersect $\overline{AB}$ at Q. Finally, draw $\overleftrightarrow{PQ}$ which is the required line.

Proof: We know that a median bisects the area of a triangle. (Why?) Thus, △ BM$_c$ has half the total area. But in terms of area, we have △ BCM$_c$ = △ BCQ + △ QCM$_c$ = △ BCQ + △ QCP = quadrilateral BCPQ as desired. The crucial step is the "sliding over" of vertex M$_c$ to point P along the line $\overleftrightarrow{PM_c}$ which is parallel to $\overleftrightarrow{CQ}$, thus △ QCM$_c$ and △ QCP have the same area, as in the previous construction.

3. Construct an equilateral triangle with a vertex on each of three parallel lines.

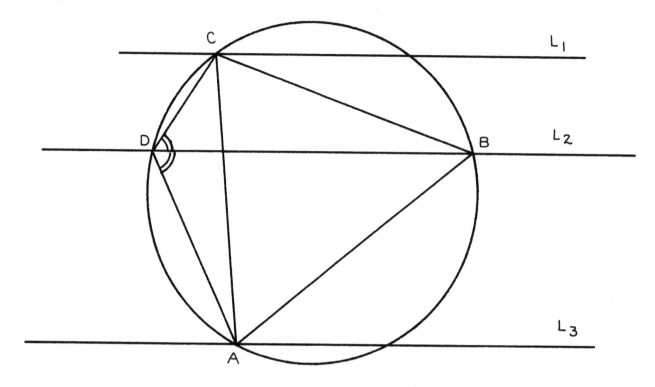

Suppose the solution, $\triangle$ ABC, is available, and we draw its circumscribed circle as shown, to cut the middle line again at some point D. Then, $\angle$ CDB and $\angle$ BDA are inscribed angles which are congruent respectively to $\angle$ BAC and $\angle$ BCA. But these angles all have measures of 60° and can be constructed. The solution starts, then, with any point D on the middle line, L_2, and the construction of angles of measure 60° above and below L_2 to give us points C and A on L_1 and L_3. Then the circumcircle of $\triangle$ CDA will cut the middle line again in vertex B of the required triangle.

Exercises

These are hard ones!

57. Construct an equilateral triangle with one vertex on each of any three lines
   (a) when they are concurrent; (b) when just two are parallel; (c) when they
   meet in three distinct points.

58. Construct an equilateral triangle with a vertex on each of three concentric
   circles.

4. Mark one point on each of the four sides of a square, then erase the square, leaving only the four points. *Problem:* Get back the square.

Another way of stating this problem is: Given, any four points; construct a square so that each side (or its exterior) goes through one of those four points.

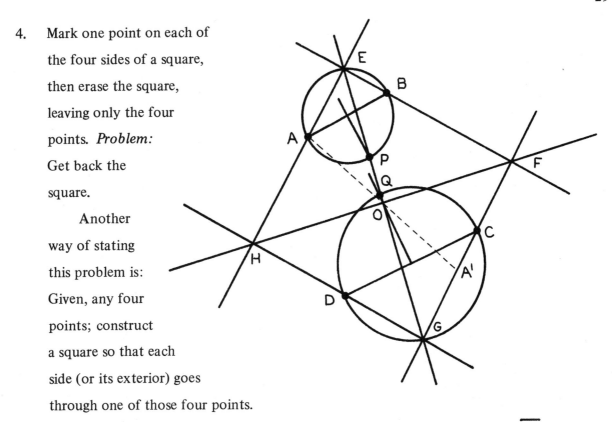

Assume that the square EFGH is available to us, and draw one diagonal, $\overline{EG}$. Since $\angle$ AEB is a right angle, if we start with given points A and B, we have an available locus for E, that is, the circle with diameter $\overline{AB}$, as shown. But the diagonal $\overline{EG}$ also bisects the opposite angles of the square and, since $\overrightarrow{EG}$ bisects $\angle$ AEB, it follows that $\overrightarrow{EG}$ must bisect the opposite arc of this circle, that is, $\overline{EG}$ must go through P, the midpoint of the inner semicircle with diameter $\overline{AB}$. For analogous reasons, $\overline{EG}$ must also go through Q, the midpoint of the inner semicircle with diameter $\overline{CD}$. But these semicircles can be constructed from the given material, and we can find the solution as follows.

Construct circles with diameters $\overline{AB}, \overline{BC}, \overline{CD}, \overline{DA}$ (only two of these are shown). Find the midpoints of a pair of opposite inside semicircles, such as P and Q, shown. Then $\overleftrightarrow{PQ}$ will contain one of the required diagonals, in this case, $\overline{EG}$, as shown. In like manner, we can get the line which contains the other diagonal, in this case $\overline{FH}$, though we do not show these construction steps. The two diagonal lines intersect at O, the center of the required square.

Once we have the center of the square, we could continue in several ways, of which this one is perhaps the simplest:

Draw $\overline{AO}$ and double that segment to A' so that $AO = OA'$. Since A and A' must lie on opposite sides of the square (why?), then A' and C must lie on the *same* side. Thus, $\overleftrightarrow{A'C}$ will meet the diagonal lines in vertices F and G of the desired square, and you should be able to finish up by yourself.

Exercises

59. Carry through the full construction as shown.

60. Discuss the situation where three or four of the given points may be collinear.

61. Place the given points so that they may lie (one or more) on the extensions of the sides as well as on their interiors. How is our analysis affected by this possibility? How is the construction affected?

5. Construct a triangle given these three points: Vertex A, the midpoint of $\overline{BC}$, the center, O, of the circumcircle. That is, construct a triangle given, in position $\{A, M_a, O\}$.

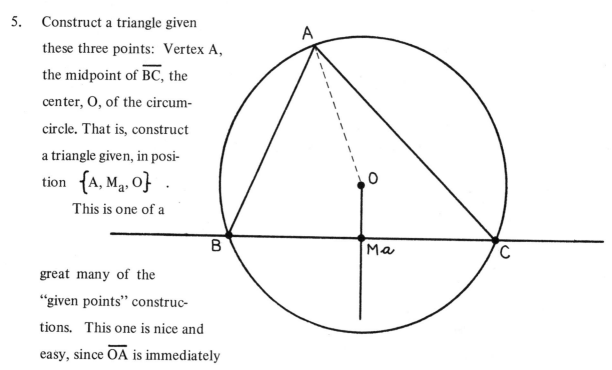

This is one of a great many of the "given points" constructions. This one is nice and easy, since $\overline{OA}$ is immediately available as the circumradius, and (O, OA) is the circumcircle. Then the chord $\overline{BC}$ will lie along the line perpendicular to $\overline{OM_a}$ at M_a, which we can draw from the given O and M_a. Finally, the circle (O, OA) will cut this line in the other two vertices B and C of the solution $\triangle ABC$.

Exercises

Construct a triangle given, in position, the points:

62. $\{ A, B, M_a \}$ (M_a is the midpoint of $\overline{BC}$.)

63. $\{ A, B, I \}$ (I is the center of the inscribed circle.)

64. $\{ A, B, T_a \}$ (T_a is the foot of the bisector of $\angle$ BAC.)

65. $\{ O, M_a, M_b \}$ (O is the center of the circumscribed circle.)

66. $\{ M_a, M_b, M_c \}$

Make up a few more of these "fixed points" construction problems. *

---

* For further discussion of such problems, see William Wernick, "Triangle Constructions with Three Located Points," *The Mathematics Magazine,* Vol. 55, No. 4 (September 1982), pp. 227-30.

Chapter III

TRIANGLE CONSTRUCTIONS

The earliest postulates on congruence of triangles are directly related to problems on the construction of triangles. For example, the postulate that two triangles are congruent if they agree in two sides and their included angle (SAS) is a direct consequence of the fact that, with the usual postulates on the uses of our instruments, a *unique* triangle can be constructed if we are given two of its sides and the angle they form. This triangle is unique in the sense that, if any other triangle were to be constructed from the same initial information, i.e., the same lengths of two sides, and the same measure of their included angle, then these two triangles would agree in all their other parts and would, that is, be congruent. In this sense, we say that this set of initial information (SAS) *determines* a unique triangle.

The figure at the top of the following page illustrates some of the details we will consider in this chapter. We list these systematically now, with the general understanding that we may use a symbol ambiguously when we can simplify our work without confusion. Thus, we may use "b" to represent either a side of a triangle, or its name, or its measure, as the context should make clear. The ambiguity reflects our choice and not our ignorance, since our aim is clarity. The rigor and precision that support the material could certainly be supplied, but only with time and space that seem inappropriate here.

Sides: a, b, c

Angles: α, β, γ

Vertices: A, B, C

Altitudes: h_a, h_b, h_c

Feet of the altitudes: H_a, H_b, H_c

Orthocenter (point of concurrence of altitudes): H

Medians: m_a, m_b, m_c

Midpoints of sides: M_a, M_b, M_c

Centroid (point of concurrence of medians): G

Angle-bisectors: t_a, t_b, t_c

Feet of angle bisectors: T_a, T_b, T_c

Incenter (point of concurrence of angle-bisectors; center of inscribed circle): I

Inradius (radius of inscribed circle): r

Circumcenter (point of concurrence of perpendicular bisectors of sides; center of circumscribed circle: O

Circumradius (radius of circumscribed circle): R

Semi-perimeter (half the sum of the lengths of the sides: $\frac{1}{2}(a + b + c)$): s

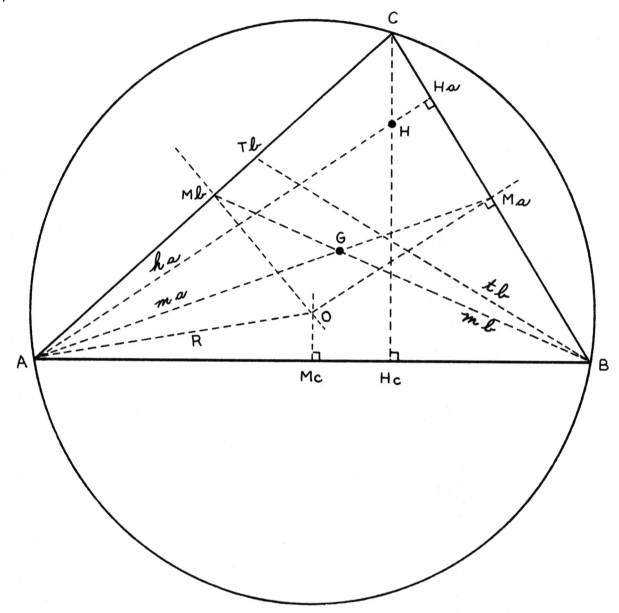

Note that we use a lower-case letter, in general, to represent a measure (of a length or angle), and a capital letter to represent a point. An exception is the use of the capital letter "R" for the length of the circumradius, to agree with this general use in the literature.

Much of the study of geometry is concerned with relations among these many items. Many of these relations are already known to you; thus, the sum of the degree measures of the three angles of a triangle is 180, i.e., $\alpha + \beta + \gamma = 180$. Many are not, thus, the reciprocal of the inradius is equal to the sum of the reciprocals of the altitudes, i.e., $\frac{1}{r} = \frac{1}{h_a} + \frac{1}{h_b} + \frac{1}{h_c}$. We will use whatever relations we may need in our constructions, with brief indications of their proofs, where appropriate. You are urged to follow through on these, to complete a proof by yourself when you can, to look up a proof in a more comprehensive reference, and possibly to find some new relations by yourself.

We consider triangle constructions on two levels: On the first level, we assume that an actual triangle exists somewhere, that someone then hands you some parts of that triangle, and your job then is to *re-construct* that original triangle. On this first level, we assume that a solution exists and that we have to find it. Our method in this case is usually to assume that the solution is available and to find appropriate relations and a sequence of steps to reconstruct the triangle.

On the second level we do not necessarily assume that a solution exists. We examine the given material only in the light of the possibility that a solution exists, and if so, to determine the relation between that given information, and the nature and number of the possible solutions. We illustrate both approaches in a familiar problem: Construct a triangle given its three sides a, b, c (below):

If we assume that someone actually had $\triangle$ ABC, then "took it apart" and gave us just the three lengths, a, b, c, we can quickly reconstruct that triangle:

On any line, take any point as the vertex A, then with arc (A, c), cut the line at B, thus making AB = c. Then draw the arcs (A, b) and (B, a)

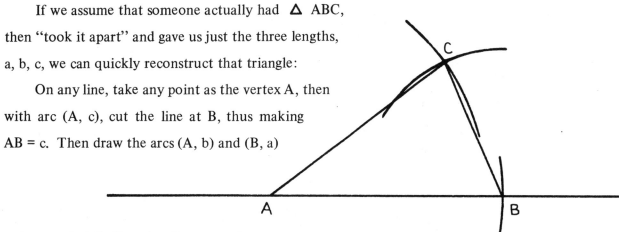

to intersect at C. Drawing the segments $\overline{AC}$ and $\overline{BC}$ will give us the required $\triangle$ ABC.

The second approach to such a problem does not assume that a solution exists, it "examines the gift horse in the mouth." If, for example, the given lengths are 2, 3 and 6, it should not take long to see that *no* triangle can be drawn with these as the lengths of its sides. Any attempt to carry through such steps as we did above will soon show the impossibility of an essential event: the intersecting of the circles (A, b) and (B, a), without which we do not have the third vertex, C. (Refer to illustration on page 36.)

We are led thus to an essential requirement in any set of three lengths which we propose as the lengths of sides of a triangle: the sum of any two of them must be greater than the third. (This is called "the triangle inequality.") If this condition is not satisfied, there can be *no* solution. If this condition *is* satisfied, then we *can* construct the triangle. As a matter of fact, the circles (A, b) and (B, a) will then intersect in two points, C and C$'$, one above, the other below the line $\overleftrightarrow{AB}$. The two triangles we thus obtain: $\triangle$ ABC and $\triangle$ ABC$'$, are symmetric to each other and are congruent, so we have essentially one solution, if any, to this problem. (What happens if a + b = c?)

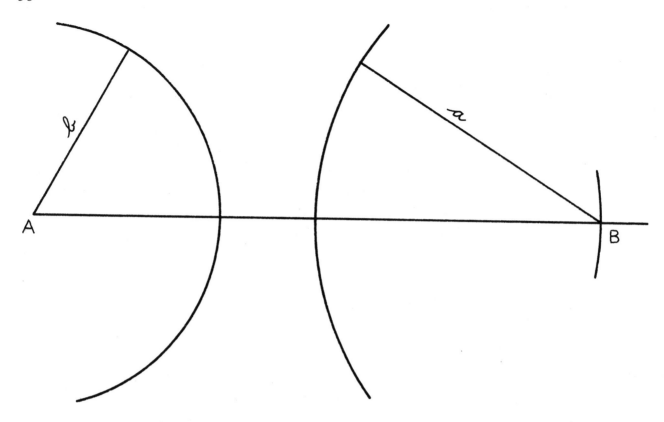

Such discussions of the possibility and numbers of solutions, and their relations to the given data can lead to deeper and more interesting mathematics. We will do some of it in later work, but whether we do or not, we urge you to pursue these ideas as often as you can.

Another difficulty may arise on both levels. Consider, for example, the problem of constructing a triangle, given its three angles $\{\alpha, \beta, \gamma\}$. If these angles came from an actual triangle, we know that the sum of their measures would have to be the measure of a straight angle; that is, any one of them would be the supplement of the sum of the other two.

Thus, if we are given any two of these angles, we need not be given the third, since we can find it ourselves from the two that are given. Such a set of information, in which some part need not be given, since it can be found from the rest, is called a *redundant set*. In this case, we are given essentially only some information about *two* angles,

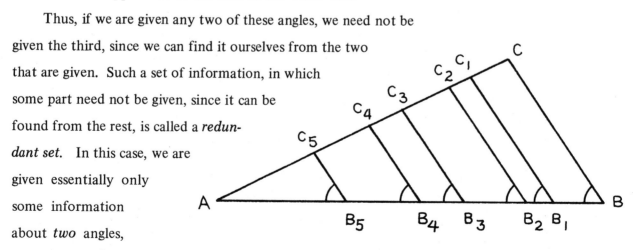

say α and β, which "actually" came from $\triangle$ ABC, but could as well have come from any one of infinitely many similar triangles, $\triangle$ AB_1C_1, $\triangle$ AB_2C_2, $\triangle$ AB_3C_3, . . .

If, on the other hand, we start with any three given angles, $\{\alpha, \beta, \gamma\}$, then they must satisfy the *necessary* condition that the sum of their measures must be the measure of a straight angle, otherwise there can be *no* triangle with these three given angles.

It should be clear that to construct or to reconstruct any particular triangle, we must have three *independent* pieces of information about it. Any dependencies among these pieces of information may make the set *redundant* and, therefore, insufficient to determine a triangle. Thus the set $\{\alpha, \beta, \gamma\}$ is redundant, since in degree measure, for example, we have $\alpha + \beta = 180 - \gamma$. Note that the set $\{a, b, c,\}$ is independent, since any choice of a and b does *not* determine c. Of course, we are bound by the triangle inequality which states, in effect, that for any solution to exist, we must have $a - b < c < a + b$. (Could you show how the first kind of inequality follows from the second kind?)

The set $\{\alpha, \beta, \gamma\}$ is quite familiar as a redundant set, but we now call attention to two other, less familiar, redundant sets.

Since a right triangle is determined when we know its hypotenuse and one acute angle, it follows that the set $\{\alpha, b, h_c\}$ is a redundant set.

In this figure, it should be clear that we can construct the right $\triangle$ ACH_c if we are given any *two* of the set $\{\alpha, b, h_c\}$, and then vertex B could be taken any-

where on $\overleftrightarrow{AH_c}$, so that we have certainly not determined any particular $\triangle$ ABC. In the right triangle, we have $h_c = b \sin \alpha$, so the set $\{\alpha, b, h_c\}$ is redundant: if we are given α and b, you need not give us h_c since we can find it ourselves. We have an analogous situation if we are given $\{a, \beta, h_c\}$ which is another redundant set (consider right $\triangle$ BCH_c), and does not determine any unique $\triangle$ ABC.

Another less obvious redundant set is (a, α, R). Suppose, as in the figure at the top of page 38, that we have drawn the circumcircle of $\triangle$ ABC, and also radii $\overline{OB}, \overline{OC}$ and the altitude $\overline{OM_a}$ of isosceles $\triangle$ OBC. From relations involving central and inscribed angles, we can see that if α is acute, then $m \angle BOC = 2m \angle \alpha$, while if α' is obtuse, then $m \angle BOC = 2m \angle (\text{supp. } \alpha')$. But from right $\triangle$ OCM_a, we have $\frac{a}{2} = R \sin \alpha$. Therefore, in both cases, $a = 2R \sin \alpha$. With this relationship, it should be clear that if we are given any two of the set $\{a, \alpha, R\}$, we can find the third ourselves, and this set is thus redundant.

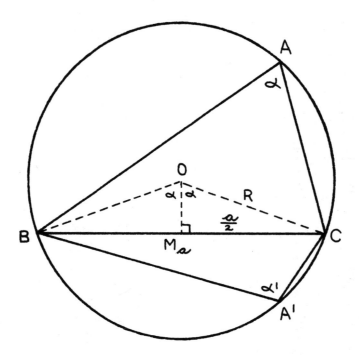

The following is a systematic and comprehensive listing of all 179 sets of three independent data, selected from the list on page 33, which may determine a triangle. Each set could have been listed in various other ways, by changing around the letters. Thus the set, 2, (a, b, α), which can be verbalized as "two sides and the angle opposite one of them," could have been represented by other choices of sides and angles, as long as we have two sides and the angle opposite one of them. For example: (a, b, β) or (a, c, α) or (a, c, γ) or (b, c, β) or (b, c, γ). If you make up or come across a problem of this type, you can find it in this listing by writing the given information in the order which we use in each set: sides, angles, altitudes, medians, angle-bisectors, circumradius, inradius and semi-perimeter.

1. $\left\{ a, b, c \right\}$

2. $\left\{ a, b, \alpha \right\}$

3. $\left\{ a, b, \gamma \right\}$

4. $\left\{ a, \alpha, \beta \right\}$

5. $\left\{ a, b, h_a \right\}$

6. $\left\{ a, b, h_c \right\}$

7. $\left\{ a, \alpha, h_a \right\}$

8. $\left\{ a, \alpha, h_b \right\}$

9. $\left\{ a, \beta, h_a \right\}$

10. $\left\{ a, \beta, h_b \right\}$

11. $\left\{ \alpha, \beta, h_a \right\}$

12. $\left\{ a, h_a, h_b \right\}$

13. $\left\{ a, h_b, h_c \right\}$

14. $\left\{ \alpha, h_a, h_b \right\}$

15. $\left\{ \alpha, h_b, h_c \right\}$

16. $\left\{ h_a, h_b, h_c \right\}$

17. $\left\{ a, b, m_a \right\}$

18. $\left\{ a, b, m_c \right\}$

19. $\left\{ a, \alpha, m_a \right\}$

20. $\left\{ a, \alpha, m_b \right\}$

21. $\left\{ a, \beta, m_a \right\}$

22. $\left\{ a, \beta, m_b \right\}$

23. $\left\{ a, \beta, m_c \right\}$

24. $\left\{ \alpha, \beta, m_a \right\}$

25. $\{a, h_a, m_a\}$

26. $\{a, h_a, m_b\}$

27. $\{a, h_b, m_a\}$

28. $\{a, h_b, m_b\}$

29. $\{a, h_b, m_c\}$

30. $\{\alpha, h_a, m_a\}$

31. $\{\alpha, h_a, m_b\}$

32. $\{\alpha, h_b, m_a\}$

33. $\{\alpha, h_b, m_b\}$

34. $\{\alpha, h_b, m_c\}$

35. $\{h_a, h_b, m_a\}$

36. $\{h_a, h_b, m_c\}$

37. $\{a, m_a, m_b\}$

38. $\{a, m_b, m_c\}$

39. $\{\alpha, m_a, m_b\}$

40. $\{\alpha, m_b, m_c\}$

41. $\{h_a, m_a, m_b\}$

42. $\{h_a, m_b, m_c\}$

43. $\{m_a, m_b, m_c\}$

44. $\{a, b, t_a\}$

45. $\{a, b, t_c\}$

46. $\{a, \alpha, t_a\}$

47. $\{a, \alpha, t_b\}$

48. $\{a, \beta, t_a\}$

49. $\{a, \beta, t_b\}$

50. $\{a, \beta, t_c\}$

51. $\{\alpha, \beta, t_a\}$

52. $\{a, h_a, t_a\}$

53. $\{a, h_a, t_b\}$

54. $\{a, h_b, t_a\}$

55. $\{a, h_b, t_b\}$

56. $\{a, h_b, t_c\}$

57. $\{\alpha, h_a, t_a\}$

58. $\{\alpha, h_a, t_b\}$

59. $\{\alpha, h_b, t_a\}$

60. $\{\alpha, h_b, t_b\}$

61. $\{\alpha, h_b, t_c\}$

62. $\{h_a, h_b, t_a\}$

63. $\{h_a, h_b, t_c\}$

64. $\{a, m_a, t_a\}$

65. $\{a, m_a, t_b\}$

66. $\{a, m_b, t_a\}$

67. $\{a, m_b, t_b\}$

68. $\{a, m_b, t_c\}$

69. $\{\alpha, m_a, t_a\}$

70. $\{\alpha, m_a, t_b\}$

71. $\{\alpha, m_b, t_a\}$

72. $\{\alpha, m_b, t_b\}$

73. $\{\alpha, m_b, t_c\}$

74. $\{h_a, m_a, t_a\}$

75. $\{h_a, m_a, t_b\}$

76. $\{h_a, m_b, t_a\}$

77. $\{h_a, m_b, t_b\}$

78. $\{h_a, m_b, t_c\}$

79. $\{m_a, m_b, t_a\}$

80. $\{m_a, m_b, t_c\}$

81. $\{a, t_a, t_b\}$

82. $\{a, t_b, t_c\}$

83. $\{\alpha, t_a, t_b\}$

84. $\{\alpha, t_b, t_c\}$

85. $\{h_a, t_a, t_b\}$

86. $\{h_a, t_b, t_c\}$

87. $\{m_a, t_a, t_b\}$

88. $\{m_a, t_b, t_c\}$

89. $\{t_a, t_b, t_c\}$

90. $\{a, b, R\}$

91. $\{a, \beta, R\}$

92. $\{\alpha, \beta, R\}$

93. $\{a, h_a, R\}$

94. $\{a, h_b, R\}$

95. $\{\alpha, h_a, R\}$

96. $\{\alpha, h_b, R\}$

97. $\{h_a, h_b, R\}$

98. $\{a, m_a, R\}$

99. $\{a, m_b, R\}$

100. $\{\alpha, m_a, R\}$

101. $\{\alpha, m_b, R\}$

102. $\{h_a, m_a, R\}$

103. $\{h_a, m_b, R\}$

104. $\{m_a, m_b, R\}$

105. $\{a, t_a, R\}$

40

106. $\{\, a,\ t_b,\ R\, \}$
107. $\{\, \propto,\ t_a,\ R\, \}$
108. $\{\, \propto,\ t_b,\ R\, \}$
109. $\{\, h_a,\ t_a,\ R\, \}$
110. $\{\, h_a,\ t_b,\ R\, \}$
111. $\{\, m_a,\ t_a,\ R\, \}$
112. $\{\, m_a,\ t_b,\ R\, \}$
113. $\{\, t_a,\ t_b,\ R\, \}$
114. $\{\, a,\ b,\ r\, \}$
115. $\{\, a,\ \propto,\ r\, \}$
116. $\{\, a,\ \beta,\ r\, \}$
117. $\{\, \propto,\ \beta,\ r\, \}$
118. $\{\, a,\ h_a,\ r\, \}$
119. $\{\, a,\ h_b,\ r\, \}$
120. $\{\, \propto,\ h_a,\ r\, \}$
121. $\{\, \propto,\ h_b,\ r\, \}$
122. $\{\, h_a,\ h_b,\ r\, \}$
123. $\{\, a,\ m_a,\ r\, \}$
124. $\{\, a,\ m_b,\ r\, \}$
125. $\{\, \propto,\ m_a,\ r\, \}$
126. $\{\, \propto,\ m_b,\ r\, \}$
127. $\{\, h_a,\ m_a,\ r\, \}$
128. $\{\, h_a,\ m_b,\ r\, \}$
129. $\{\, m_a,\ m_b,\ r\, \}$
130. $\{\, a,\ t_a,\ r\, \}$

131. $\{\, a,\ t_b,\ r\, \}$
132. $\{\, \propto,\ t_a,\ r\, \}$
133. $\{\, \propto,\ t_b,\ r\, \}$
134. $\{\, h_a,\ t_a,\ r\, \}$
135. $\{\, h_a,\ t_b,\ r\, \}$
136. $\{\, m_a,\ t_a,\ r\, \}$
137. $\{\, m_a,\ t_b,\ r\, \}$
138. $\{\, t_a,\ t_b,\ r\, \}$
139. $\{\, a,\ R,\ r\, \}$
140. $\{\, \propto,\ R,\ r\, \}$
141. $\{\, h_a,\ R,\ r\, \}$
142. $\{\, m_a,\ R,\ r\, \}$
143. $\{\, t_a,\ R,\ r\, \}$
144. $\{\, a,\ b,\ s\, \}$
145. $\{\, a,\ \propto,\ s\, \}$
146. $\{\, a,\ \beta,\ s\, \}$
147. $\{\, \propto,\ \beta,\ s\, \}$
148. $\{\, a,\ h_a,\ s\, \}$
149. $\{\, a,\ h_b,\ s\, \}$
150. $\{\, \propto,\ h_a,\ s\, \}$
151. $\{\, \propto,\ h_b,\ s\, \}$
152. $\{\, h_a,\ h_b,\ s\, \}$
153. $\{\, a,\ m_a,\ s\, \}$
154. $\{\, a,\ m_b,\ s\, \}$

155. $\{\, \propto,\ m_a,\ s\, \}$
156. $\{\, \propto,\ m_b,\ s\, \}$
157. $\{\, h_a,\ m_a,\ s\, \}$
158. $\{\, h_a,\ m_b,\ s\, \}$
159. $\{\, m_a,\ m_b,\ s\, \}$
160. $\{\, a,\ t_a,\ s\, \}$
161. $\{\, a,\ t_b,\ s\, \}$
162. $\{\, \propto,\ t_a,\ s\, \}$
163. $\{\, \propto,\ t_b,\ s\, \}$
164. $\{\, h_a,\ t_a,\ s\, \}$
165. $\{\, h_a,\ t_b,\ s\, \}$
166. $\{\, m_a,\ t_a,\ s\, \}$
167. $\{\, m_a,\ t_b,\ s\, \}$
168. $\{\, t_a,\ t_b,\ s\, \}$
169. $\{\, a,\ R,\ s\, \}$
170. $\{\, \propto,\ R,\ s\, \}$
171. $\{\, h_a,\ R,\ s\, \}$
172. $\{\, m_a,\ R,\ s\, \}$
173. $\{\, t_a,\ R,\ s\, \}$
174. $\{\, a,\ r,\ s\, \}$
175. $\{\, \propto,\ r,\ s\, \}$
176. $\{\, h_a,\ r,\ s\, \}$
177. $\{\, m_a,\ r,\ s\, \}$
178. $\{\, t_a,\ r,\ s\, \}$
179. $\{\, R,\ r,\ s\, \}$

This list is to be considered as a list of 179 construction *problems,* which you are invited to plunge into. In the rest of this chapter, we will do a few for you to show you some useful techniques or to develop some geometric information you may not have come across before. In the first few, we will discuss rather fully the possibility and number of solutions under various conditions, but in later problems, we will leave this interesting (and more difficult) work for you.

Selected Constructions.

5. $\{a, b, h_a\}$

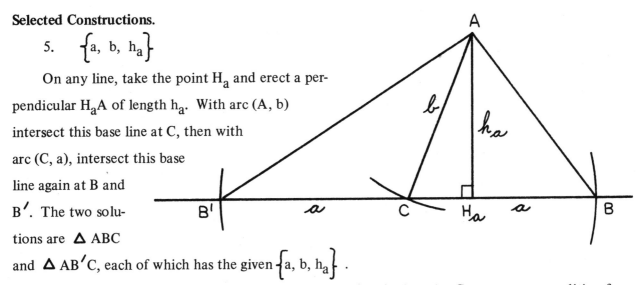

On any line, take the point H_a and erect a perpendicular H_aA of length h_a. With arc (A, b) intersect this base line at C, then with arc (C, a), intersect this base line again at B and B'. The two solutions are $\triangle ABC$ and $\triangle AB'C$, each of which has the given $\{a, b, h_a\}$.

Discussion: Since arc (A, b) must cut the base line to obtain point C, a necessary condition for a solution is that $b \geq h_a$. Since this arc will cut the line again (at C' to the right of H_a), we will also have another pair of solutions, but these will be reflections of the ones we already have. Another possibility would follow if we had taken the perpendicular at H_a below as well as above the base line, but again, we have reflections of the solutions we found before, and we will not in later work discuss such reflections or symmetric solutions that contribute nothing essentially new. If $b = h_a$, then there will be just one point of contact between (A, b) and the base line, at the point H_a itself, which is then a point of tangency of this arc. In this case, the two triangles $\triangle ACB$ and $\triangle ACB'$ become congruent right triangles, that is, we have essentially a single solution; but in other possible cases, with $b > h_a$, we get two solutions, no matter what length is chosen for a. Thus, finally, the condition $b \geq h_a$ is necessary and sufficient for any solution to this problem, with the equality leading to one solution, and the inequality leading to two.

7. $\{a, \propto, h_a\}$

This problem is nicely done by intersection of loci as explained in Chapter II.

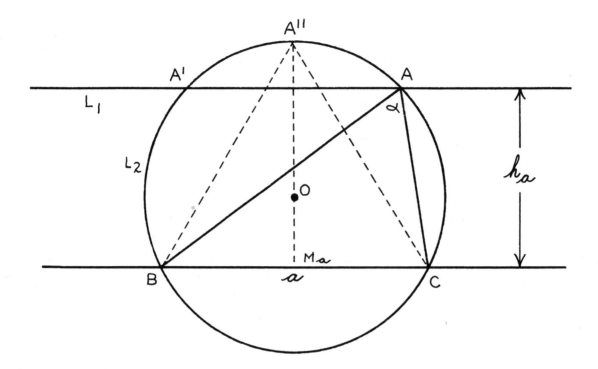

On any line make BC = a. Then one locus for vertex A is a line parallel to $\overleftrightarrow{BC}$ at distance h_a, i.e., L_1, since every point on L_1 is at distance h_a from this base line. Another locus for vertex A is the circular arc which subtends $\overline{BC}$, and in which any inscribed angle would have measure equal to α . Thus, to solve this problem, we construct L_1 and L_2 as indicated, to intersect at A, then draw $\overline{AB}$ and $\overline{AC}$ to finish the solution $\triangle$ ABC. (Another solution, congruent to this one, would come from A', the other intersection of the loci.)

Discussion: We will have solutions if and only if the two loci intersect, which will occur if h_a is not "too big." Consider the "tallest" triangle, $\triangle$ A″ BC, and in particular, right $\triangle$ A″BM$_a$, in which m $\angle$ BA″M$_a$ = ½ α and BM$_a$ = ½a. The original problem will have a solution if and only if $h_a \leq$ A″M$_a$ which becomes, from relations in that right triangle: $h_a \leq$ ½a cot ½ α . With equality here, we will have L_1 tangent to L_2 at A″ and the only solution will be the isosceles $\triangle$ A″BC; with inequality, we will have the two congruent triangles, $\triangle$ ABC and $\triangle$ A′ BC, that is, essentially one solution.

13. $\left\{ a,\ h_b,\ h_c \right\}$

On any line, make BC = a, and construct a semicircle on $\overline{BC}$ as diameter. (See the figure on page 43.) This semicircle is a locus for both H$_b$ and H$_c$, since both $\angle$ BH$_b$C and $\angle$ BH$_c$C are right angles. Now, draw (B, h_b) to intersect this semi-circle at H$_b$, and then (C, h_c) to intersect it at H$_c$. Finally, $\overleftrightarrow{BH_c}$ and $\overleftrightarrow{CH_b}$ intersect at A to give us the solution $\triangle$ ABC.

We leave the full discussion to you, with a hint that you should examine the relative lengths, a, h_a, h_c, which determine the intersections of the various arcs that enter into the construction.

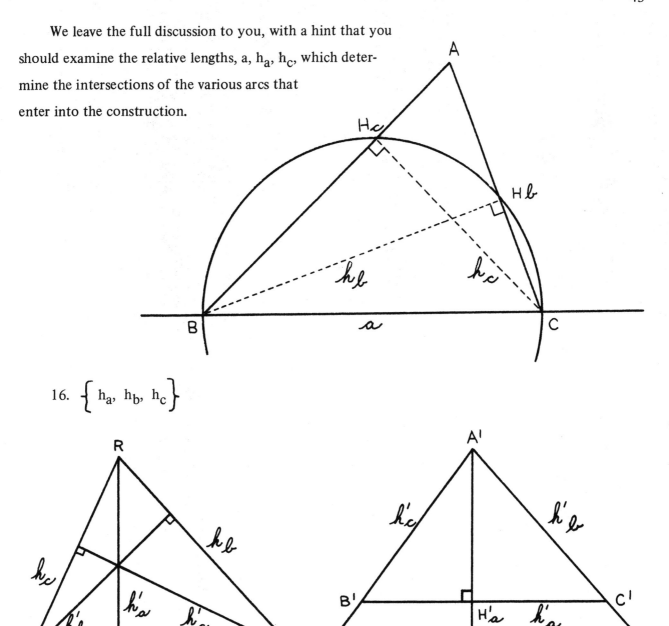

16. $\left\{ h_a, h_b, h_c \right\}$

Solution I: Since the area of △ ABC is ½a h_a, which equals ½bh_b, which equals ½ch_c, we have: $ah_a = bh_b = ch_c$ from which we could write: $a : \frac{1}{h_a} = b : \frac{1}{h_b} = c : \frac{1}{h_c}$. These last equations tell us that the sides of a triangle are inversely proportional to their corresponding altitudes, and conversely, since we also have $h_a : \frac{1}{a} = h_b : \frac{1}{b} = h_c : \frac{1}{c}$. Then, if we make a new △ PQR, with sides, h_a, h_b and h_c, then the altitudes of *this* △ PQR, i.e., h_a', h_b', and h_c', will also be inversely proportional to *its* sides, h_a, h_b and h_c, i.e., $h_a : \frac{1}{h_a'} = h_b : \frac{1}{h_b'} = h_c : \frac{1}{h_c'}$. But now we can see that these new altitudes will be directly proportional to the sides of the original $\triangle$ ABC, i.e., $a : h_a' = b : h_b' = c : h_c'$. Therefore, a new triangle whose sides are h_a', h_b' and h_c' will be similar to the required $\triangle$ ABC.

The construction is then: (1) construct $\triangle$ PQR whose sides are the given altitudes, h_a, h_b and h_c, (2) find the altitudes h_a', h_b', h_c' of $\triangle$ PQR; (3) construct $\triangle$ A′B′C′ whose sides are these altitudes just found (this $\triangle$ A′B′C′ is similar to the solution $\triangle$ ABC); (4) construct any altitude, say $\overline{A'H_a'}$ of $\triangle$ A′B′C′, and on $\overline{A'H_a'}$, lay off $\overline{A'H_a}$ to be congruent to the given altitude of length h_a; (5) through H_a, draw a perpendicular to cut $\overleftrightarrow{A'B'}$ and $\overleftrightarrow{A'C'}$ in B and C. Then, $\triangle$ A′BC is the required triangle.

Solution II: If secants are drawn from an outside point E to a circle, then

EF • EG = EK • EL = EM • EN,

and so on. These equal products can be related to the equal products $ah_a =$ $bh_b = ch_c$ with the following

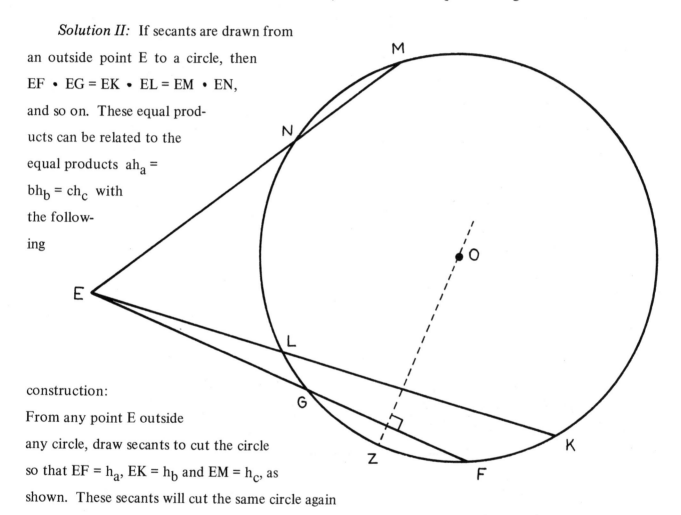

construction:

From any point E outside any circle, draw secants to cut the circle so that $EF = h_a$, $EK = h_b$ and $EM = h_c$, as shown. These secants will cut the same circle again at G, L and N, respectively, and we name the corresponding lengths $EG = a'$, $EL = b'$ and $EN = c'$. Since the construction leads to $a'h_a = b'h_b = c'h_c$, we obtain by division: $a : a' = b : b' = c : c'$. Thus, a triangle whose sides have the lengths a', b', c', just found, will be similar to the solution $\triangle$ ABC. We continue the construction by making $\triangle$ A′B′C′ with sides congruent to $\overline{EG}$, $\overline{EL}$ and $\overline{EN}$, then proceed as in Solution I, since in both cases, $\triangle$ A′B′C′ will be similar to the solution $\triangle$ ABC.

20. $\{a, \alpha, m_b\}$

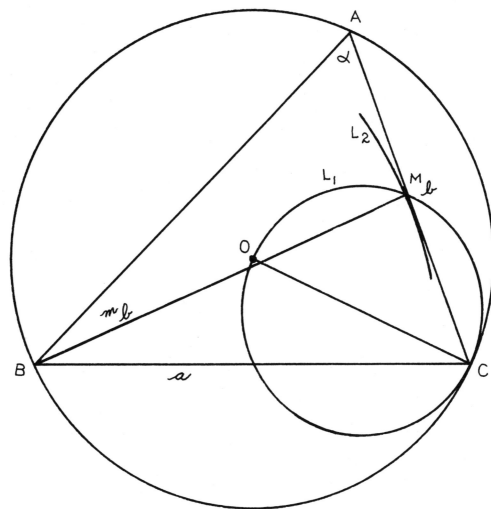

On any line, lay off BC = a, and then construct the circular arc $\overset{\frown}{BAC}$ to contain α. This arc is a locus for the vertex A, and its circle is the circumcircle of $\triangle$ ABC. Then construct the circle with diameter $\overline{OC}$. This circle, L_1, is a locus for the midpoints of all arcs of the first circle that can be drawn from C, and is thus a locus for M_b. Since the distance from B to M_b is the given median length m_b, another locus for M_b is L_2, the circle (B, m_b) which we draw. Thus, M_b is at the intersection of these loci, as shown, and $\overleftrightarrow{CM_b}$ will meet the first circle at A. Then, $\triangle$ ABC is the solution.

Exercise

1. Could we take M_b as the other intersection of the two loci above? Explain.

34. $\left\{a,\ h_b,\ m_c\right\}$

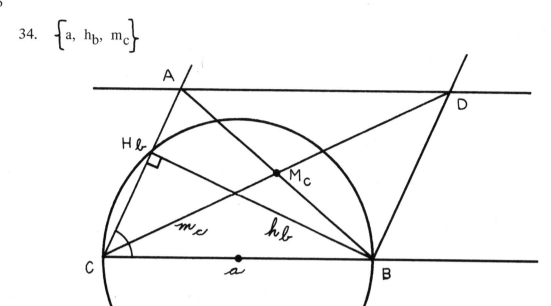

Suppose the triangle is constructed. If we double $\overline{CM_c}$ to D, then quadrilateral ACBD is a parallelogram, since diagonals bisect each other. The lengths a and h_b determine the right $\triangle\ BH_bC$, and thus the parallelogram ACBD. Hence the construction:

On any line, make CB = a, and then construct a semicircle on diameter $\overline{BC}$. (This semicircle is a locus for H_b.) With arc $(B,\ h_b)$, cut this semicircle at H_b, then draw $\overleftrightarrow{CH_b}$, then draw $\overleftrightarrow{BD}$ through B and parallel to $\overleftrightarrow{CH_b}$. With arc $(C,\ 2m_c)$, cut $\overleftrightarrow{BD}$ at D, which is a third vertex of the parallelogram. Through D, a line parallel to $\overleftrightarrow{BC}$ is drawn to intersect $\overleftrightarrow{CH_b}$ in A, the fourth vertex of the parallelogram, and the third vertex of our solution, $\triangle$ ABC.

Exercises

2. The arc $(C,\ 2m_c)$ may cut $\overleftrightarrow{BD}$ at another point, D$'$. Discuss the construction and the solution you get from there on.

3. If a has length 10 inches, what are allowable lengths for h_b? for m_c? Discuss.

35. $\left\{h_a,\ h_b,\ m_a\right\}$

Suppose the triangle is available. Then, in right $\triangle$ AM$_a$H$_a$, we know the hypotenuse, m_a, and leg, h_a, so that the triangle is determined. In right $\triangle$ AM$_a$K, since M$_a$ is the midpoint of side $\overline{BC}$, its distance to side $\overline{AC}$ will be half the distance from B to $\overline{AC}$; i.e., M$_a$K = ½h_b, thus in this right $\triangle$ Am$_a$K, we also know its hypotenuse length, m_a, and the length of leg $\overline{M_aK}$ = ½h_b. Hence the construction:

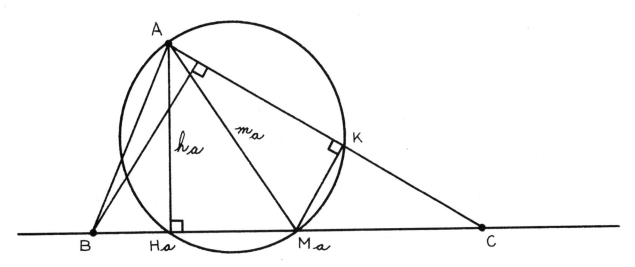

On any line, make $AM_a = m_a$, then draw circle with $\overline{AM_a}$ as diameter. This circle is a locus for both H_a and K. Cut this circle with (A, h_a) to locate H_a, then with (M_a, ½h_b) to locate K. Then, $\overleftrightarrow{AK}$ and $\overleftrightarrow{H_aM_a}$ intersect at vertex C. Finally, double $\overline{CM_a}$ to find B and draw $\overline{AB}$ to complete the solution $\triangle$ ABC.

Exercises

4. Suppose we took K and H_a on the same side of $\overline{AM_a}$, rather than on opposite sides, as in our figure. Complete the figure and discuss.

5. Under what conditions would $\overleftrightarrow{AK}$ and $\overleftrightarrow{H_aM_a}$ fail to meet? How would our "solution" be affected in that case?

6. If m_a = 10 inches, discuss the possible lengths of h_a and h_b for any solution or for any number of solutions.

43. $\left\{ m_a,\ m_b,\ m_c \right\}$

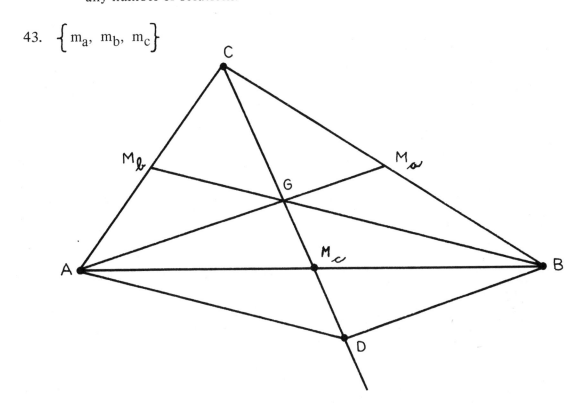

Suppose the triangle is available. Double $\overline{GM_c}$ to D and draw $\overline{AD}$ and $\overline{DB}$. Then, ADBG is a parallelogram since its diagonals, $\overline{AB}$ and $\overline{GD}$, bisect each other. But AG = $\frac{2}{3}$ m_a, AD = BG = $\frac{2}{3}$ m_b, and GD = 2(GM$_c$) = $\frac{2}{3}$ m_c. Thus, $\triangle$ ADG has sides whose lengths are respectively 2/3rds the lengths of the given medians. Hence the construction:

Construct, as in Chapter I, segments whose lengths are 2/3rds of the given medians, and with these segments, construct $\triangle$ ADG, with sides, say AD = $\frac{2}{3}$ m_b, DG = $\frac{2}{3}$ m_c and GA = $\frac{2}{3}$ m_a. Now, construct the median $\overline{AM_c}$ of this triangle, and double its length to B, a vertex of the required triangle. Finally, double the length of $\overline{DG}$ to C and draw sides $\overline{AC}$ and $\overline{BC}$ of the solution $\triangle$ ABC.

Exercises

7. What conditions must the lengths m_a, m_b, m_c satisfy if we are to have any solution?

8. What are the consequences if any two or all three of the given medians have equal lengths?

56. $\left\{ a, h_b, t_c \right\}$

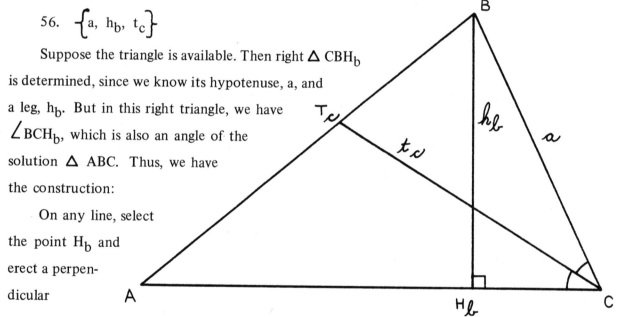

Suppose the triangle is available. Then right $\triangle$ CBH$_b$ is determined, since we know its hypotenuse, a, and a leg, h_b. But in this right triangle, we have $\angle$BCH$_b$, which is also an angle of the solution $\triangle$ ABC. Thus, we have the construction:

On any line, select the point H$_b$ and erect a perpendicular

there, making H$_b$B = h_b. Then arc (B, a) will cut this base line at C, and we can draw $\overline{BC}$. Then bisect $\angle$BCH$_b$ and on this bisector make CT$_c$ = t_c. Finally, the intersection of $\overleftrightarrow{BT_c}$ and $\overleftrightarrow{CH_b}$ is the vertex A of solution $\triangle$ ABC.

Exercises

9. What conditions on a and h_b would make it impossible to construct right $\triangle$BCH$_b$?

10. What length of the angle bisector $\overline{CT_c}$ would make it impossible for lines $\overleftrightarrow{BT_c}$ and $\overleftrightarrow{CH_b}$ to meet?

11. What conditions among $\left\{ a, h_b, t_c \right\}$ would lead to an isosceles triangle? to an equilateral triangle?

63. $\{h_a, h_b, t_c\}$

Suppose the triangle is available. From T_c, draw $\overline{T_cY}$ perpendicular to $\overline{BC}$. Now, since an angle bisector divides the opposite side into segments whose lengths are proportional to the lengths of the adjacent sides, we have: $\dfrac{AT_c}{T_cB} = \dfrac{b}{a}$. Also, since the lengths of the sides of a triangle are inversely proportional to the lengths of the corresponding altitudes (see Solution 16, this chapter), we have $\dfrac{b}{a} = \dfrac{h_a}{h_b}$. Combining these equations, we have $\dfrac{AT_c}{T_cB} = \dfrac{h_a}{h_b}$. From this proportion, we obtain (how?)

$$\frac{h_a + h_b}{h_b} = \frac{AT_c + T_cB}{T_cB} = \frac{AB}{T_cB}.$$

But, from right $\triangle BAH_a$ and right $\triangle BT_cY$, we also have: $\dfrac{AB}{T_cB} = \dfrac{AH_a}{T_cY} = \dfrac{h_a}{T_cY}$. Therefore, finally, $\dfrac{h_a + h_b}{h_b} = \dfrac{h_a}{T_cY}$. This shows T_cY as fourth proportional to known quantities, and therefore constructible, as in Chapter I.

The construction starts, then, with finding the segment $\overline{T_cY}$ as fourth proportional, from the available lengths, $h_a + h_b$, h_b and h_a. Then we construct right $\triangle CYT_c$, knowing its hypotenuse $CT_c = t_c$ and leg $\overline{T_cY}$. But this right triangle contains $\angle YCT_c$, which has half the measure of $\angle BCA$ of our solution triangle. If we copy $\angle YCT_c$ on the other side of $\overline{CT_c}$, then the ray $\overrightarrow{CA}$ is one locus L_1 for vertex A. But since this vertex is also at distance h_a from its opposite side, another locus for A is L_2, the line parallel to $\overleftrightarrow{CY}$ and distance h_a from it. These two loci intersect at vertex A, and finally the lines $\overleftrightarrow{AT_c}$ and $\overleftrightarrow{CY}$ intersect at B, the third vertex of our solution $\triangle ABC$.

This solution came through an algebraic analysis that was far from obvious. Of course it was necessary to know the geometric relations that led to the proportions. Our later constructions may lean on even more unfamiliar geometric relations, and you are urged to deepen and extend your knowledge of geometry as you venture into the rougher waters that lie ahead.

74. $\left\{h_a, m_a, t_a\right\}$

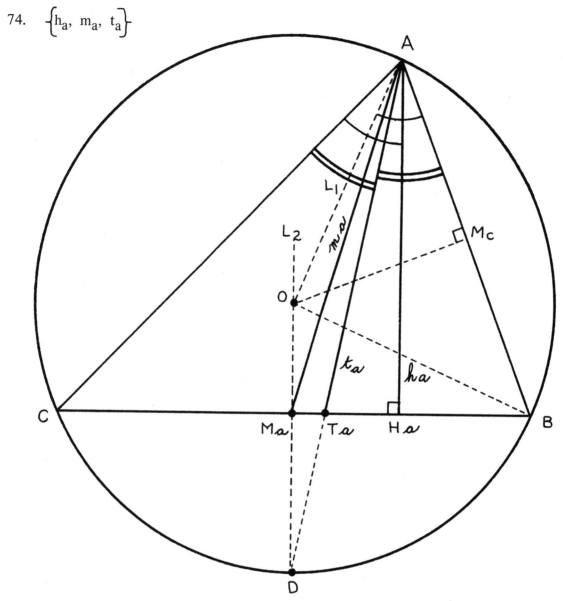

Suppose, as usual, that the solution triangle is available. Then the right $\triangle AH_aT_a$ and the right $\triangle AH_aM_a$ are both determined by the fact that in each case we have a known leg and hypotenuse. If we draw the circumcircle and then radius $\overline{OA}$ and radius $\overline{OM_aD}$, we can prove here a minor theorem of essential importance in this conclusion: *An angle bisector of a triangle also bisects the angle formed by the altitude and circumradius from the same vertex.*

Proof: (Outline only, you supply details.) The perpendicular from O to $\overline{AB}$ meets $\overline{AB}$ at M_c, the midpoint of $\overline{AB}$. Since central angle $\angle AOB$ and inscribed angle $\angle ACB$ intercept the same arc, we have $\gamma = \frac{1}{2} m \angle AOB = m \angle AOM_c$. Therefore, from right $\triangle ACH_a$ and right $\triangle OAM_c$, we have $m \angle CAH_a = m \angle OAM_c$ = complement of γ. But, since $\overrightarrow{AT_a}$ bisects $\angle BAC$, we have $m \angle BAT_a = m \angle CAT_a$. Therefore, by subtraction, we have $m \angle OAT_a = m \angle H_aAT_a$; thus, $\overrightarrow{AT_a}$ bisects not only $\angle BAC$, but also $\angle H_aAO$, as we said.

The same figure leads to another useful conclusion: *An angle bisector of a triangle meets the perpendicular bisector of its opposite side on the circumcircle of that triangle.* That is, $\overleftrightarrow{OM_a}$ and $\overleftrightarrow{AT_a}$ meet at D, on the circumcircle. This result follows immediately from the fact that both of these lines must bisect the arc $\overset{\frown}{BC}$.

Our construction follows readily from the first of these two theorems: Construct right $\triangle AH_aT_a$ and right $\triangle AH_aM_a$ as usual, then double $\angle H_aAT_a$ beyond $\overrightarrow{AT_a}$ to obtain one locus L_1 for the circumcenter O (since $\overrightarrow{AT_a}$ bisects $\angle H_aAO$). Another locus for O is the perpendicular, L_2, to $\overleftrightarrow{H_aM_a}$ at M_a. Finally, the circle (O, OA) will meet $\overleftrightarrow{H_aM_a}$ at B and C, the other two vertices of the solution $\triangle ABC$.

Discussion: Of course we must have both m_a and t_a greater than h_a in order to construct our first two right triangles.

Exercises

12. Discuss the situations in which there is any equality among the three given lengths.

13. We took M_a and T_a on the same side of H_a. Discuss the consequences of taking them on opposite sides of H_a.

14. In what circumstances would the circle (O, OA) fail to meet $\overleftrightarrow{H_aM_a}$?

99. $\{a,\ m_b,\ R\}$

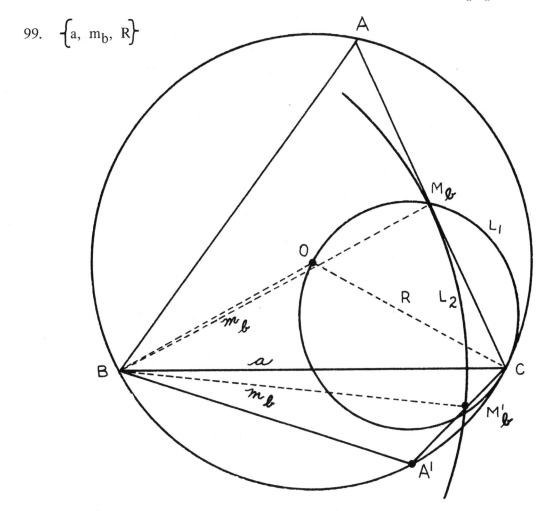

Construct the isosceles $\triangle OBC$ with known sides: $OB = OC = R$, $BC = a$. Then the circle (O, R) is a locus for the third vertex of the solution $\triangle ABC$. But since M_b is the midpoint of chord $\overline{AC}$, one locus for M_b is the circle L_1 with $\overline{OC}$ as diameter. Since M_b is at a known distance, m_b, from B, another locus for M_b is L_2, the circle (B, m_b) which will intersect L_1 at M_b and at M'_b. Then, $\overleftrightarrow{CM_b}$ and $\overleftrightarrow{CM'_b}$ will meet the circumcircle (O, R) at the third vertices, A, A', of the solution triangles $\triangle ABC$, $\triangle A'BC$.

We have shown in this case the two distinct solutions obtained from the given data. Of course, the given lengths could have led to loci that might not intersect at all, in which case there would have been no solution.

Exercises

15. Discuss the number and possibility of solutions with different selections of the given lengths $\{a, m_b, R\}$.

16. See if you can arrive at the following necessary and sufficient conditions for any solution: $\sqrt{R^2 + 2a^2} - R \leq 2m_b \leq \sqrt{R^2 + 2a^2} + R$.

102. $\{h_a, m_a, R\}$

Suppose the solution is available. Then in right $\triangle AH_aM_a$, we know the length of hypotenuse $AM_a = m_a$ and the length of leg $AH_a = h_a$, so this triangle is determined. But the circumcenter can then be found, because it is on the perpendicular bisector of $\overline{BC}$, i.e., on the perpendicular to $\overleftrightarrow{M_aH_a}$ at M_a, and it's also at known distance R from vertex A. Hence the construction:

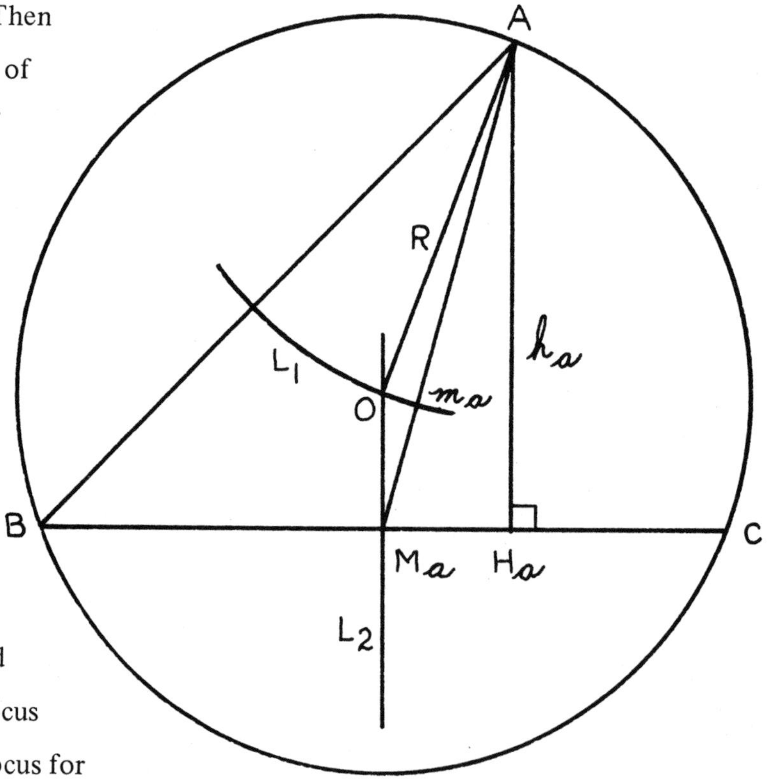

Construct right $\triangle AH_aM_a$ with known hypotenuse length $AM_a = m_a$ and known leg length $AH_a = h_a$. Then one locus for O is L_1, the circle (A, R). Another locus for O is L_2, the perpendicular to $\overleftrightarrow{H_aM_a}$ at M_a. These intersect at the circumcenter O. Finally, with (O, R), we intersect the base line $\overleftrightarrow{H_aM_a}$ at B and C, the other two vertices of the solution $\triangle ABC$.

Exercises

17. Under what circumstances would it be impossible to draw the first right △ AH$_a$M$_a$?

18. Under what circumstances would the loci L$_1$ and L$_2$ fail to intersect?

19. These loci might intersect twice: at O, as shown, and at O$'$, not shown. Follow through on this possibility.

20. Once we have the intersection O, could we ever fail to get the last two vertices B and C? Discuss.

105. $\{$a, t$_a$, R$\}$

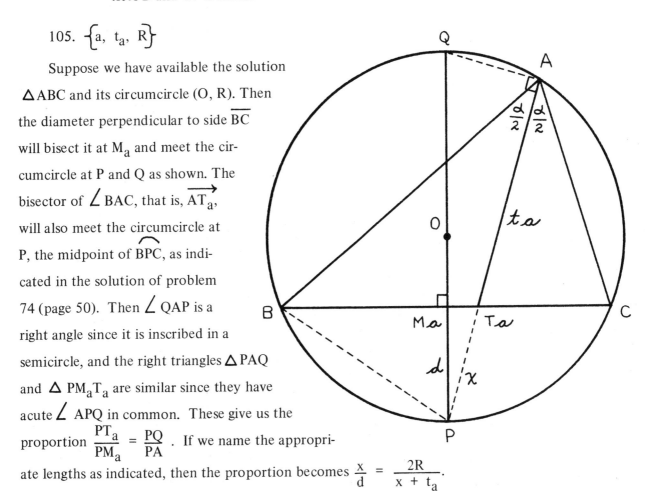

Suppose we have available the solution △ABC and its circumcircle (O, R). Then the diameter perpendicular to side $\overline{BC}$ will bisect it at M$_a$ and meet the circumcircle at P and Q as shown. The bisector of ∠ BAC, that is, $\overrightarrow{AT_a}$, will also meet the circumcircle at P, the midpoint of $\overset{\frown}{BPC}$, as indicated in the solution of problem 74 (page 50). Then ∠ QAP is a right angle since it is inscribed in a semicircle, and the right triangles △ PAQ and △ PM$_a$T$_a$ are similar since they have acute ∠ APQ in common. These give us the proportion $\dfrac{PT_a}{PM_a} = \dfrac{PQ}{PA}$. If we name the appropriate lengths as indicated, then the proportion becomes $\dfrac{x}{d} = \dfrac{2R}{x + t_a}$.

Since $\overline{PQ}$ is the circumdiameter, its length, 2R, is known and of course, t$_a$ is given. The length d is readily found, once we have drawn the chord of length a in the circumcircle of radius R, since d is exactly the distance from the midpoint of that chord to the midpoint of its arc. (Could you show that d = R $-$ $\sqrt{R^2 - a^2/4}$?)

Thus the proportion leads to an equation in which x is to be found in terms of available lengths, d, 2R and t$_a$: x(x + t$_a$) = d(2R).

We show now a geometric solution to such equations: $x(x + u) = vw$, where u, v and w are known.

On any line, lay off $\overline{JK}$ of length v and $\overline{JL}$ of length w, then draw $\overline{LM}$ of length u in any direction. Circumscribe $\triangle$ KLM with circle π_1 and then find distance, z, from its center to $\overline{LM}$. Now draw circle (Y, z) and draw a tangent to this circle from out-

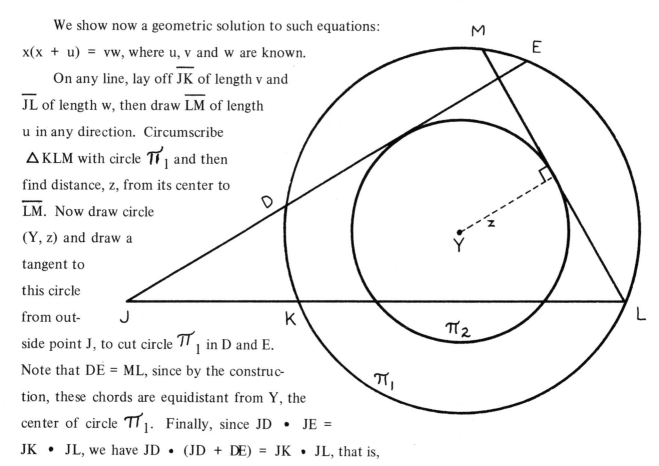

side point J, to cut circle π_1 in D and E. Note that DE = ML, since by the construction, these chords are equidistant from Y, the center of circle π_1. Finally, since JD • JE = JK • JL, we have JD • (JD + DE) = JK • JL, that is, JD • (JD + u) = vw, and we have thus found JD = x, such that $x(x + u) = vw$, as wanted.

Now, to put these steps together for our actual construction, we start by drawing the circumcircle (O, R) and in it placing the chord $\overline{BC}$ of given length a. The perpendicular bisector of $\overline{BC}$ gives us the lengths $d = PM_a$ and 2R = PQ for the next construction steps, in a separate figure.

On any line, lay off $JK = PM_a = d$ and JL = PQ = 2R, then from L in any direction, draw $\overline{LM}$ of given length t_a. Circumscribe $\triangle$ KLM as described, then draw the perpendicular from the center Y to $\overline{LM}$, then draw circle π_2, that is concentric with π_1 and tangent to $\overline{LM}$. Draw, as in Chapter I, a tangent line $\overleftrightarrow{JDE}$ to circle π_2, and from the theory above, we now have JD = x.

Back in our first figure, the circle (P, x) is now drawn to intersect $\overline{BC}$ at point T_a, and finally $\overleftrightarrow{PT_a}$ will meet the circumcircle at A, the third vertex of the solution $\triangle$ ABC.

Discussion: We have gone through a lot of good algebra and geometry and we leave further details and comments to you, noting only that in the original figure, the four points A, Q, M_a and T_a all lie on a circle (with diameter $\overline{QT_a}$), since both $\angle T_aAQ$ and $\angle T_aM_aQ$ are right angles.

115. $\{a, \alpha, r\}$

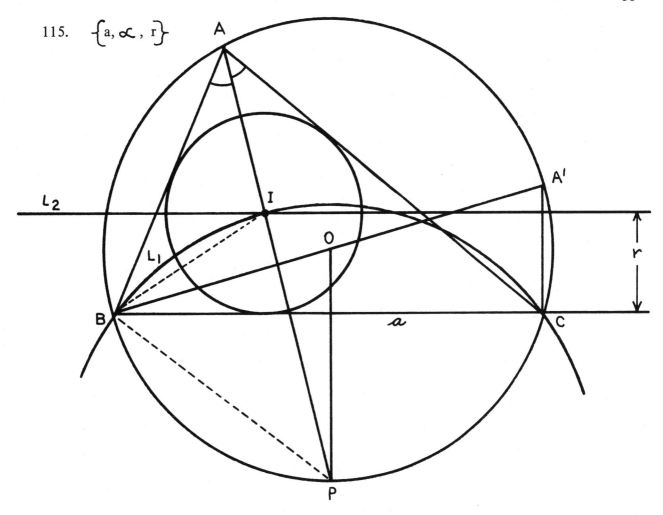

This solution will also lead us through some good geometry which may not be familiar to you. Suppose, as usual, that the solution triangle is available. We know that the bisector of $\angle$ BAC goes through the incenter I and the midpoint P, of the opposite arc $\overset{\frown}{BPC}$ of the circumscribed circle. Consider $\triangle$ PBI: Since $\angle$ BPI and $\angle$ BCA are inscribed angles, both of which intersect $\overset{\frown}{AB}$, therefore m $\angle$ BPI = m $\angle$ BCA = γ . Also, m $\angle$ PBI = m $\angle$ PBC + m $\angle$ CBI = $\frac{\alpha}{2}$ + $\frac{\beta}{2}$ since $\overrightarrow{BI}$ and $\overrightarrow{AI}$ are both angle bisectors of the solution $\triangle$ ABC. Thus, since $\alpha + \beta + \gamma$ = 180°, we find that m $\angle$ BIP = $\frac{\alpha}{2}$ + $\frac{\beta}{2}$, and therefore $\triangle$ PBI is isosceles, with $\overline{PB} \cong \overline{PI}$. But $\overline{PB}$ can be found from the given information, since a and α are enough to determine the circumcircle, as indicated at the beginning of this chapter, and once we have placed the known chord $\overline{BC}$ in the known circumcircle (O, OB), we can easily draw radius $\overline{OP}$ as the perpendicular bisector of that chord, and then draw $\overline{PB}$.

Thus we have one locus, L_1 for I, the circle (P, PB) which is available from the given information. Another locus for I follows from the fact that the incircle is tangent to each side and its center, I, is thus r units distant from each side. Thus, our second locus for I is the line L_2, parallel to $\overleftrightarrow{BC}$ and r units above it.

The construction follows this analysis: Construct right $\triangle$ BCA$'$, with BC = a, right angle at C and m $\angle$ CBA$'$ = complement of α, then circumscribe this triangle. This circle (O, OB) is the circumcircle of the solution triangle, because any angle inscribed in $\overparen{BA'C}$ will have angle measure α, as arranged for $\angle$ BA$'$C. Draw radius $\overline{OP}$ perpendicular to chord $\overline{BC}$, then draw the circle (P, PB) which is our first locus L_1 for the incenter I. Now draw the second locus, L_2, which is a line parallel to $\overleftrightarrow{BC}$ and r units above it. These two loci meet at the incenter I, and then $\overleftrightarrow{PI}$ meets the circumcircle at A, the third vertex of solution $\triangle$ ABC.

Discussion: Of course, L_1 and L_2 must meet if we are to locate I, and if they meet once, they may meet again. We leave further discussion about the number and nature of the solutions to you.

122. $\left\{h_a,\ h_b,\ r\right\}$

We will not solve this problem completely here, but we will indicate how we can reduce it to a problem we have already solved. We develop first a pleasant little theorem which we mentioned earlier in this chapter.

Consider first the solution $\triangle$ ABC, with its incircle (I, r) and angle bisectors $\overrightarrow{AI}$, $\overrightarrow{BI}$ and $\overrightarrow{CI}$. Then the area of $\triangle$ ABC may be found by adding the areas of $\triangle$IAB, $\triangle$ IBC and $\triangle$ ICA, whose bases have lengths respectively c, a and b, and whose common altitude has length r. Thus:

Area $\triangle$ ABC = ½ra + ½rb + ½rc = ½r(a + b + c), therefore: $a + b + c = \dfrac{2(\text{Area } \triangle \text{ ABC})}{r}$.

But from much earlier work, we know that: Area $\triangle$ ABC = ½ah$_a$ = ½bh$_b$ = ½ch$_c$, thus:

$a = \dfrac{2(\text{Area } \triangle \text{ ABC})}{h_a}$, $b = \dfrac{2(\text{Area } \triangle \text{ABC})}{h_b}$ and $c = \dfrac{2(\text{Area } \triangle \text{ ABC})}{h_c}$. If we add corresponding

terms in these last three equations and then combine with the earlier one, we are led to the result:

$\dfrac{1}{r} = \dfrac{1}{h_a} + \dfrac{1}{h_b} + \dfrac{1}{h_c}$ which may be expressed in words thus: "The reciprocal of the inradius is

equal to the sum of the reciprocals of the altitudes" — certainly a neat and unexpected result.

The algebraic consequence of this equation is, given any three of these four quantities, we can find the fourth. Since this problem starts with h_a, h_b and r as given, it has now been reduced, theoretically at least, to Problem 16, in which we were given the three altitudes.

There still remains the question of constructing the reciprocal of any given length x. We require a unit length, PT, and any circle tangent to $\overline{PT}$ at T. (See the figure at the top of the following page.) Let the circle (P, x) cut this circle at Q, and consider R, the other interesection of $\overleftrightarrow{PQ}$ with the first circle.

Since PQ • PR = PT^2 = 1, it follows that PQ and PR
are reciprocals. If (P, x) does not cut the first circle,
just take any larger circle tangent to $\overline{PT}$ at T,
and proceed as before.

The actual solution construction will
only be indicated now. Use this reciprocal
construction to find reciprocals of r, h_a, h_b.
Then subtract to find the reciprocal of h_c:

$$\frac{1}{h_c} = \frac{1}{r} - \frac{1}{h_a} - \frac{1}{h_b}$$

Then find the recip-
rocal of this recipro-
cal to obtain h_c it-
self, and *now* we

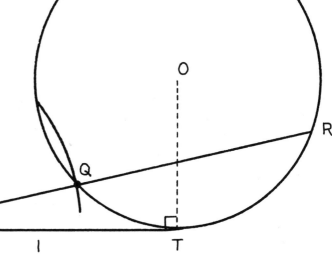

are right back into Problem 16 which was done long ago. (Refer back to page 43.)

150. $\left\{ \alpha,\ h_a,\ s \right\}$

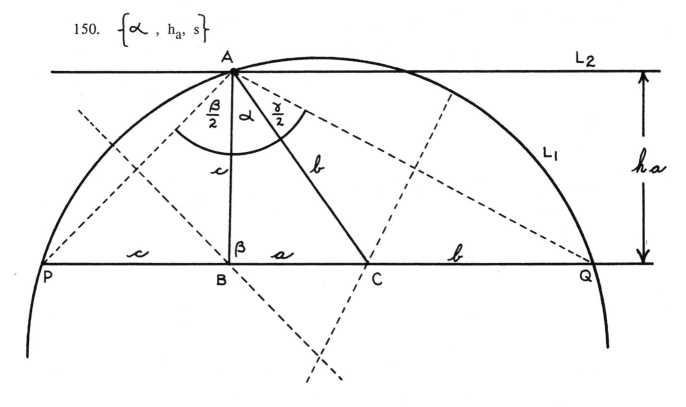

Suppose the solution triangle is available. On $\overleftrightarrow{BC}$, make BP = BA = c, CQ = CA = b, thus the
length of $\overline{PBCQ}$ is a + b + c = 2s which is known. In the isosceles $\triangle$ BPA, the measure of each of
the congruent base angles is half the measure of the exterior angle at vertex B, i.e., m $\angle$ PAB = ½ β .

Analogously, m $\angle$ QAC = ½ γ , thus at the vertex A, we have m $\angle$ PAQ = ½ β + α + ½ γ = (½ α + ½ β + ½ γ) + ½ α = 90° + ½ α , and is thus also known in terms of the original given material.

Thus, on the known segment, $\overline{PQ}$, the point A subtends a known angle 90° + ½ α , and the locus of A is thus known to be a circular arc, L_1. Another locus for A is the line L_2, parallel to $\overleftrightarrow{BC}$ and the distance h_a above it.

We start with the actual construction by finding L_1, which we will review now:

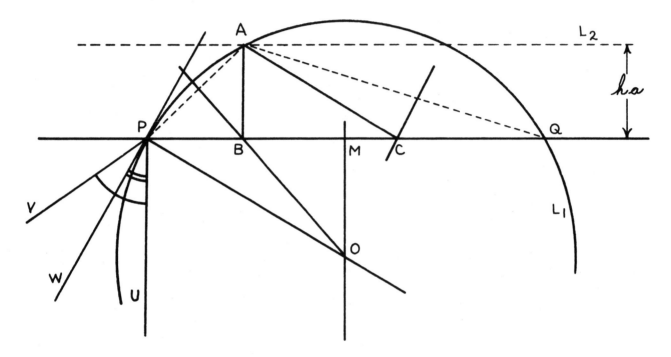

On any line, make PQ = 2s which is the perimeter of the solution $\triangle$ ABC. At P, construct $\overleftrightarrow{PU}$ perpendicular to $\overleftrightarrow{PQ}$, then make $\angle$ UPV congruent to the given angle of measure α . Bisect $\angle$ UPV with $\overrightarrow{PW}$, thus: m $\angle$ QPW = 90° + ½ α , as in the analysis. Then O, the center of arc L_1, will be found at the intersection of the perpendicular bisector of $\overline{PQ}$ and the perpendicular to $\overleftrightarrow{PW}$ at P. Finally, (O, OP) is the actual locus L_1.

The other locus, L_2, is easily drawn as indicated, and the two loci intersect at vertex A of the solution triangle. Then, at last, the perpendicular bisectors of $\overline{AP}$ and $\overline{AQ}$ meet the base line in B and C, the other two vertices of the solution $\triangle$ ABC.

We have solved only a few of the 179 problems on our list, but have selected those which lead to relevant and, we hope, interesting geometric material. We urge you to explore this territory in much more detail, because it is rewarding in knowledge and satisfaction and pleasure.

Our last exercise in this chapter is a big one!

Exercises

21. Complete the solutions of all the rest of the 179 problems, pages 38 to 40.

22. Discuss for each the conditions for the possibility of any solution and the relations among conditions for the nature and number of any solutions.

Chapter IV

CIRCLE CONSTRUCTIONS

We have used circles in many of our previous constructions, but in this chapter we investigate in detail some problems which involve the construction of a circle to fit given conditions. One familiar situation is the construction of a circle through the three vertices of a triangle, i.e., the circumscribed circle of a given triangle. The solution is well known and always uniquely possible: the perpendicular bisectors of the sides are concurrent at the circumcenter O, and the required circle has center O and radius $\overline{OA}$ (or $\overline{OB}$ or $\overline{OC}$).

Another familiar situation is the construction of the inscribed circle of a given triangle, i.e., the circle tangent to its three sides. This solution is also well known and always uniquely possible: the three angle bisectors are concurrent at the incenter I, which is the center of the desired circle. Its radius is simply the distance from I to any of the three sides of the original triangle.

Both of these problems involve the construction of a circle to go through given points or to be tangent to given lines. A natural generalization would be the problem of constructing a circle to go through one or more points *and* tangent to one or more lines, *and* perhaps tangent to one or more circles.

This larger general problem is sometimes called "The Problem of Apollonius" and is analyzed in exactly ten situations, as follows:

(1)	PPP		(6)	PLC
(2)	PPL		(7)	LLC
(3)	PLL		(8)	PCC
(4)	LLL		(9)	LCC
(5)	PPC		(10)	CCC

(1)   PPP

We have already discussed this case, when the three points are the vertices of a triangle. The only special situation to be considered is the one in which the three points may not form a triangle, that is, in which they are collinear. In this event, the only "circle" to go through them is a "circle" of infinite radius, that is, a straight line.

(2)   PPL

If the solution were available we would see that the line containing chord $\overline{P_1P_2}$ meets the given line at a point, A, which is thus an external point from which a tangent and secant are drawn to a circle. But we know in such a situation that the tangent length is the mean proportional

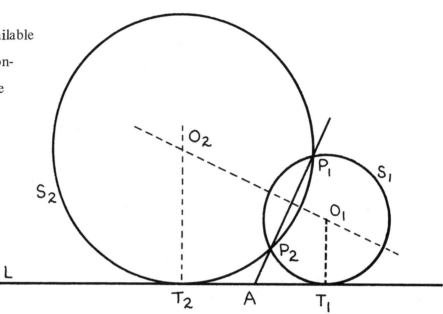

between the length of the whole secant and the length of its external segment.

Let $\overleftrightarrow{P_1P_2}$ intersect line L at A. Find, by Construction 1-14, the mean proportion, t, between AP_1 and AP_2, and mark along L, on both sides of A, the points T_1 and T_2 so that $AT_1 = AT_2 = t$. Thus, T_1 and T_2 are tangent points of the required circles on line L. The centers of these circles are on *both* the perpendicular bisector of $\overline{P_1P_2}$ *and* the perpendicular to L at T_1 and T_2.

Exercises

Certain situations may arise and should now be considered:

1. Discuss the situation if $\overleftrightarrow{P_1P_2}$ is parallel to L. (One solution.)
2. Discuss the situation if P_1 is on L_1. (One solution.)
3. Discuss the situation if P_1 and P_2 are both on L.
4. Discuss the situation if L comes between P_1 and P_2.

(3)   PLL

If the solution were available, we see that there will be, in general, two solutions with two common

points, P and P'. These points must be symmetrically placed with respect to the angle bisector $\overrightarrow{OB}$. Furthermore, the line containing common chord $\overline{PP'}$ must meet one of the sides, say L_2, at A. Since we can find $\overrightarrow{OB}$, a bisector of one of the angles formed by the given lines L_1 and L_2, and we can thus also find P', the point symmetric to P with respect to $\overleftrightarrow{OB}$, we have thus reduced the case of PLL to the case PPL which we have just discussed.

Exercises

Certain special cases may come up and should be discussed in detail. Discuss each of the following:

5. L_1 and L_2 are parallel and P lies on L_1.

6. L_1 and L_2 are parallel and P lies between them.

7. L_1 and L_2 are parallel and P lies outside of them.

8. L_1 and L_2 intersect, and P lies on L_1.

9. L_1 and L_2 intersect at P.

(4)   LLL

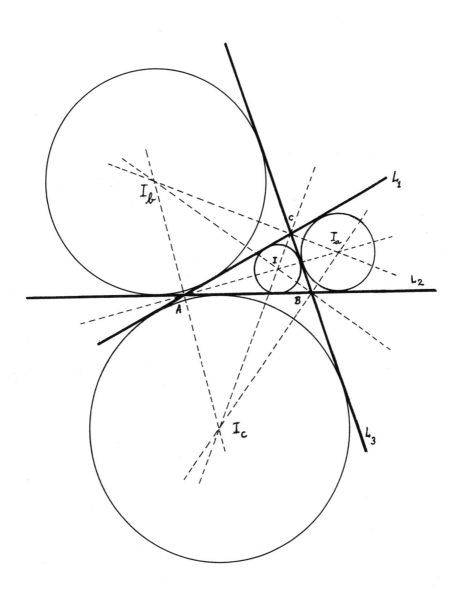

We have, in Construction 1-17, discussed briefly just one simple case — the inscribed circle of the triangle formed by three lines which meet in three distinct points, A, B, C. The figure at the bottom of page 63 shows the situation that develops when we consider the three lines that form $\triangle$ ABC. The constructions are simple enough: since each circle is tangent to three lines, its center must lie on the bisectors of the angles formed by these lines. Once we have the centers, we can easily find the radii (how?) and then draw the required circles. We urge you to study this figure carefully, it contains a lot of mathematics.

Exercises

These exercises indicate some details which you are asked to work out yourself. Try to find (and prove) other relations.

10. The circle with center I is called the *inscribed* circle of $\triangle$ ABC, and has radius r. The circles with centers I_a, I_b, I_c are called the *escribed* circles of $\triangle$ ABC, with radii r_a, r_b, r_c. They are all related by this remarkable formula:

$$\frac{1}{r} = \frac{1}{r_a} + \frac{1}{r_b} + \frac{1}{r_c}$$

Prove it. (Hint: Use areas.)

11. These four radii are also related to $\triangle$ ABC in other ways: each can be found directly from the lengths of the sides. Thus, if we represent as usual the semiperimeter s = ½(a + b + c), we have:

$$r = \sqrt{\frac{(s-a)(s-b)(s-c)}{s}}$$

$$r_a = \sqrt{\frac{s(s-b)(s-c)}{(s-a)}}; \quad r_b = \sqrt{\frac{s(s-a)(s-c)}{(s-b)}}; \quad r_c = \sqrt{\frac{s(s-a)(s-b)}{(s-c)}}$$

Prove these. (Hint: Use areas, particularly Heron's formula for the area of

$$\triangle \text{ABC} = \sqrt{s(s-a)(s-b)(s-c)}.)^*$$

12. Prove: The reciprocal of the inradius is equal to the sum of the reciprocals of the lengths of the altitudes, i.e., $\frac{1}{r} = \frac{1}{h_a} + \frac{1}{h_b} + \frac{1}{h_c}$.

13. (a) Prove: The product of all four of these radii is equal to the square of the area of the triangle.

    (b) Prove: The sum of the three exradii is equal to the sum of the inradius and four times the circumradius.

(The figure itself shows some remarkable properties involving collinearity and perpendicularity.)

14. Prove that each vertex of the triangle is collinear with the incenter, I, and its opposite excenter; thus, A, I, I_a are collinear.

---

* For a proof of Heron's formula, see Alfred S. Posamentier and Charles T. Salkind, *Challenging Problems in Geometry* (Palo Alto, Calif.: Dale Seymour Publications, 1988), pp. 135-7.

15. Prove that each vertex is also collinear with its two adjacent excenters, thus, A, I_b, I_c are collinear.

16. Prove that in $\triangle\, I_a I_b I_c$ the altitudes are exactly the bisectors of the angles of $\triangle$ABC, which meet at the incenter, I.

17. In view of the result of problem 16 above, the four points I, I_a, I_b, I_c form an orthic quadrilateral, which means that if we select any three of them and draw the altitudes of that triangle, then those altitudes will be concurrent at the fourth point. Prove this.

(5) · PPC

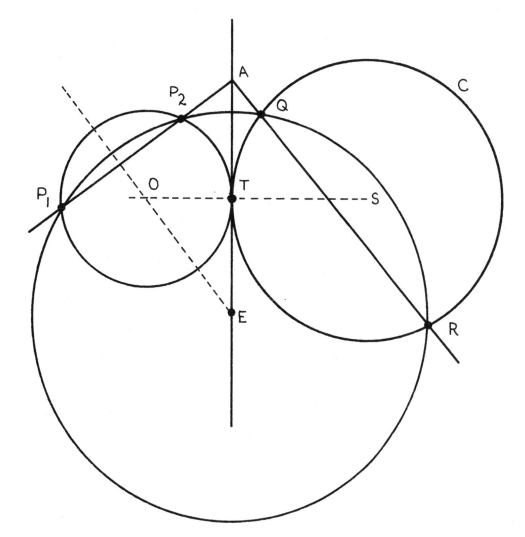

Suppose the solution were available. Then the circles would be tangent at point T, at which point we could draw a common tangent line. If, from any point A in this tangent line, we draw secants to cut the desired circle in P_1 and P_2 and the given circle in Q and R, we would have $AP_1 \cdot AP_2 = AT^2 = AR \cdot AQ$. Thus, since $AP_1 \cdot AP_2 = AR \cdot AQ$, the four points P_1, P_2, Q and R are cyclic. (See Exercise 18, which follows.) But we can easily draw a circle through given points P_1 and P_2 that will intersect given circle C. Following is the construction.

First draw the perpendicular bisector of $\overline{P_1P_2}$ (this is a locus for centers of all circles through P_1 and P_2). On this perpendicular, take any point E and draw the circle (E, EP_1) to cut given circle C in Q and R. Draw $\overleftrightarrow{QR}$ to intersect $\overleftrightarrow{P_1P_2}$ at A. From A, draw tangent line $\overleftrightarrow{AT}$ to the given circle, by Construction 1-19. Then draw $\overleftrightarrow{ST}$ through center S of the given circle and point T just found, to intersect the perpendicular bisector of $\overline{P_1P_2}$ at O, the center of the required circle. (The tangent line $\overleftrightarrow{AT}$ to the given circle is only one of the two possible tangent lines. Discuss the construction which uses the other tangent line, $\overline{AT'}$, not drawn in this figure.)

Exercises

18. If, as indicated, $\overleftrightarrow{P_1P_2}$ and $\overleftrightarrow{QR}$ intersect at A, and $AP_1 \cdot AP_2 = AR \cdot AQ$, prove that P_1, P_2, Q and R are cyclic. (Hint: Obtain a proportion from the given equation, then prove that a pair of triangles are similar, then get a pair of angles supplementary, then use the fact that a quadrilateral can be inscribed in a circle if and only if a pair of its opposite angles are supplementary.)

19. Prove: If two circles are tangent, their line of centers contains their common point of tangency.

(6)   PLC

If the solution were available, it could appear in the figure as (E, EG), tangent to the given circle C at T, to the given line L at G, and passing through the given point P. The line of centers, $\overleftrightarrow{OE}$, must go through T. (Why?) Draw the perpendicular from center O to

line L, cutting the given circle at A and B, the ends of a diameter; then draw the perpendicular from center E to L, meeting it at G which is a point of tangency. (Why?) Then draw $\overline{BT}, \overline{AT}, \overline{TG}$ and finally $\overline{AP}$ to cut the desired circle at H. Lines $\overleftrightarrow{OF}$ and $\overleftrightarrow{EG}$ are parallel, and triangles $\triangle OAT$ and $\triangle TEG$ are isosceles, with congruent vertex angles at O and E. Hence their base angles are congruent and in particular, $\angle OAT \cong \angle ETG$, hence $\overline{AT}$ and $\overline{TG}$ lie on the same straight line. Hence, $\angle BTG \cong \angle BTA \cong$ a right angle. Since $\angle BFG$ is also a right angle, we now know that BFGT is a cyclic quadrilateral (on diameter $\overline{BG}$). Therefore, we have secants from outside point A, hence $AB \cdot AF = AT \cdot AG$.

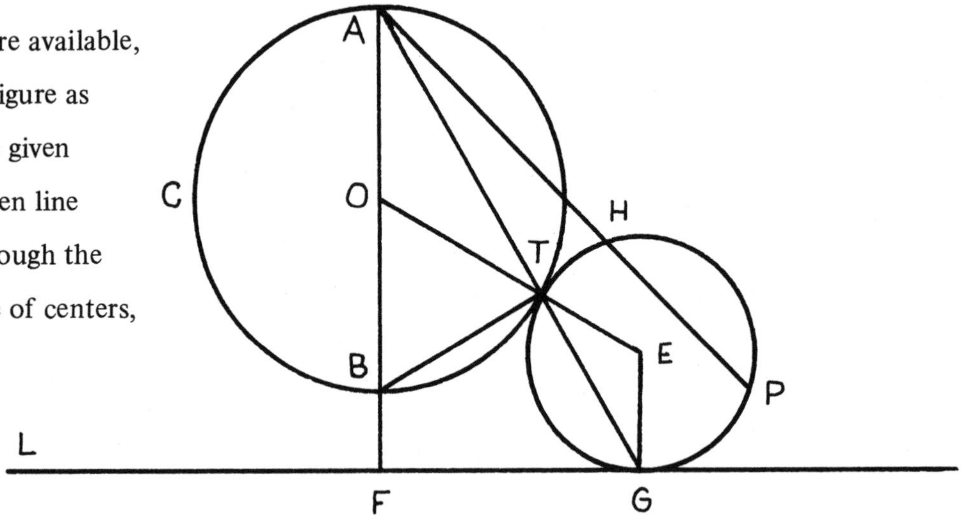

But points T, G, P and H are also cyclic and, as before, AT • AG = AH • AP; therefore AB • AF = AH • AP. But A, B, F and P are all quickly available from the given material, so we can find point H. The problem has thus been reduced to PPL, already done as problem 2.

Briefly, to find H, we draw the perpendicular from O, center of given circle C, to L, cutting C at A and B and meeting L at F. Draw $\overleftrightarrow{AP}$ and on it, construct $\overline{AH}$, found from $\dfrac{AP}{AF} = \dfrac{AB}{AH}$ (Construction 1-13). Now proceed with P, H, L, as with P_1P_2L in problem 2.

(7)   LLC

Suppose, as usual, that a solution S is available as shown in the drawing on page 68. Since circle S is tangent to given circle C, the length of segment $\overline{OA}$ joining their centers is equal to the sum of their radii, x unknown, and r known and given. Thus, another circle S$'$, concentric with S and with radius x + r will go through the center A of given circle C and be tangent to lines L_1' and L_2', parallel respectively to the given lines L_1' and L_2', and distance r beyond them. But since these lines are easily constructed, the problem has been reduced to problem 3 (PLL), with P the center A of given circle C, and with the two lines as L_1' and L_2', parallel to the given lines L_1 and L_2, but distance r beyond them. The solution S to this problem will give us the desired center O. Since we know, from problem 3, that there are in general two solutions, we should have two solutions here also, S and S* as sketched.

Another possibility must still be investigated: suppose the solution circle T is *internally* tangent to the given circle. In that case, the length of the segment $\overline{OA}$ joining their centers will be equal to the *difference* of their radii rather than the sum, as before. Now a new auxiliary circle T$'$, concentric with T but inside it, and with radius x$'$ – r will go through the center A of the given circle and be tangent to two auxiliary lines, L_1'' and L_2'', parallel respectively to given lines L_1 and L_2, but *inside* the angle rather than beyond the angle as before. We have again reduced the problem to PLL, with point A, the lines L_1'' and L_2'', as already indicated. There are two solutions in this case also, with only one of them, T, shown.

Exercises

As usual, special situations call for special treatment. Discuss each of the following, with figures and construction.

20. L_1 and L_2 are parallel and C is tangent to both.

21. L_1 and L_2 are parallel and C intersects both.

22. L_1 and L_2 are parallel and C intersects L_1 and is tangent to L_2.

23. L_1 and L_2 are parallel and C intersects L_1 but not L_2.

24. L_1 and L_2 are parallel and C lies between them, tangent to L_1 but not to L_2

25. L_1 and L_2 are parallel and C lies between them but not tangent to either.

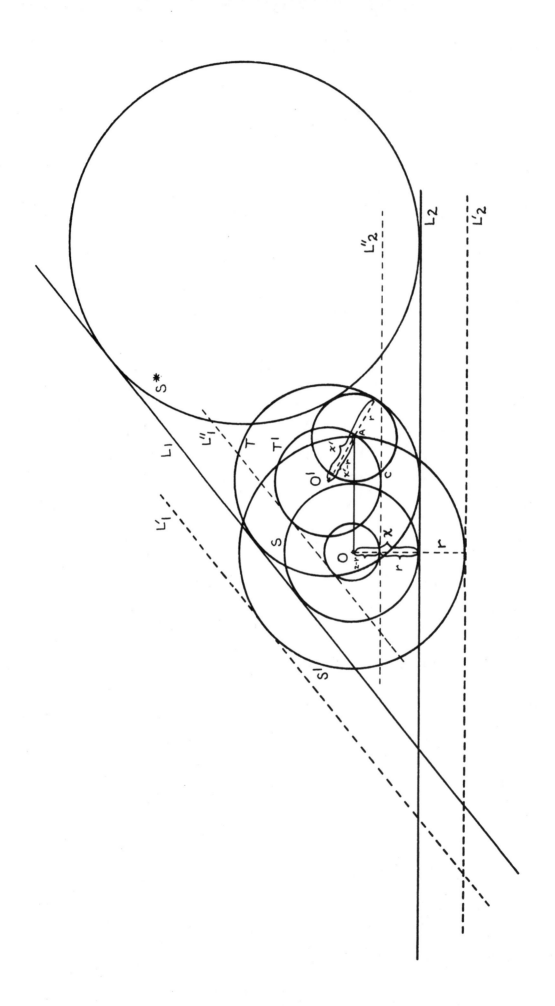

26. L_1 and L_2 intersect at point K which is interior to C.

27. L_1 and L_2 intersect at K, which is on C.

28. L_1 and L_2 intersect at K and C is tangent to L_1 at K.

29. L_1 and L_2 intersect at K outside C, but C is tangent to L_1 and L_2.

30. L_1 and L_2 intersect at K outside C, but C is tangent to L_1 and intersects L_2.

31. L_1 and L_2 intersect at K outside C, but C intersects both L_1 and L_2.

(8)   PCC

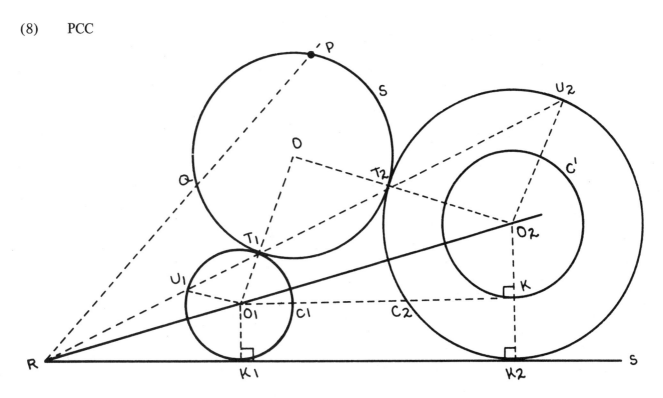

Suppose that a solution, S, is available, tangent externally to both given circles C_1 and C_2 at points T_1 and T_2. Let the common tangent $\overleftrightarrow{K_1K_2}$ to the given circles meet the line of centers, $\overrightarrow{O_1O_2}$, at point R. This is a center of similitude of these two circles. Assume (which is true, though we will not prove it here) that $\overleftrightarrow{T_1T_2}$ will go through R and draw the other lines as indicated. Then triangles $\triangle\ U_1O_1T_1$, $\triangle\ T_1OT_2$ and $T_2O_2U_2$ are all isosceles and *all* their base angles are congruent. Thus, $\overleftrightarrow{O_1U_1}\ //\ \overleftrightarrow{OO_2}$ and $\overleftrightarrow{OO_1}\ //\ \overleftrightarrow{O_2U_2}$. Thus, from two pairs of similar triangles, we have: $\dfrac{RU_1}{RT_2} = \dfrac{RO_1}{RO_2} = \dfrac{RT_1}{RU_2}$. Therefore, $RU_1 \cdot RU_2 = RT_1 \cdot RT_2$. But for each circle separately, we have $RU_1 \cdot RT_1 = RK_1{}^2$ and $RU_2 \cdot RT_2 = RK_2{}^2$, thus $(RU_1 \cdot RT_1) \cdot (RU_2 \cdot RT_2) = RK_1{}^2 \cdot RK_2{}^2 = (RU_1 \cdot RU_2) \cdot (RT_1 \cdot RT_2) = (RT_1 \cdot RT_2)^2$. Therefore, $RT_1 \cdot RT_2 = RK_1 \cdot RK_2$. This means that the four points T_1, T_2, K_2, K_1, are cyclic (as are the four points T_1, T_2, P, Q) as indicated in Exercise 18. Thus, $RQ \cdot RP = RT_1 \cdot RT_2 = RK_1 \cdot RK_2$.

70

This last product is known since it is obtainable directly from the given circles as soon as we construct their common external tangent. (To construct this tangent, we start by drawing the circle $C' = (O_2, O_2K)$, where O_2K is equal to the difference of the radii of the given circles; then drawing a tangent line from O_1 to this circle C', by Construction 1-19. Then the common tangent to the two circles will be parallel to this tangent line $\overleftrightarrow{O_1K}$ and distance O_1K_1 beyond it.)

Then, from $RQ \cdot RP = RK_1 \cdot RK_2$, we locate Q in $\overleftrightarrow{RP}$ by a fourth proportional construction, Construction 1-13, and have thus reduced the problem to Case 5 (PPC), with points P, Q, and either of the given circles.

Since S could have been drawn internally tangent to either or both of the given circles, and since P might be in a variety of locations relative to the given circles, which might in turn be in a variety of positions relative to each other, there are already many specific situations to investigate. We leave the pleasant details to you.

(9)   LLC

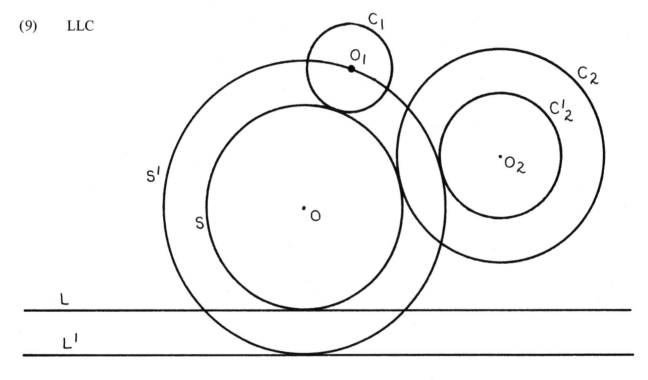

We suppose, as usual, that a solution is available, as in the figure. Given: circle C_1 and circle C_2 with radii r_1 and r_2, and line L. If we draw the circle S', concentric with the solution circle S, but with radius $\overline{OO_1}$, then we have "expanded" the solution circle S to S', which will now pass through O_1, be tangent to a new circle C'_2, concentric with C_2 and with radii equal to the *difference* of the given radii, and be tangent to a new line L', parallel to L, and beyond it by the distance r_1. But clearly we can construct circle C'_2 and line L', thus we can construct circle S' as a solution of problem 6 (PLC) with given point O_1, given line L' and given circle C'. Once we have circle S', it's easy to get the solution circle S.

Exercises

As usual, we investigate special situations in detail. Discuss each of the following, with figures and constructions.

32. C_1 is inside C_2 (not tangent) and L intersects them both.

33. C_1 is inside C_2 (not tangent) and L is tangent to C_1.

34. C_1 is inside C_2 (not tangent) and L intersects C_2 but not C_1.

35. C_1 is inside C_2 (not tangent) and L is tangent to C_2.

36. C_1 is inside C_2 (not tangent) and L intersects neither circle.

37. If C_1 is internally tangent to C_2 at T, discuss the solutions for each of the possible positions of L relative to these circles, as in Exercises 32-36.

38. As in Exercise 37, suppose C_1 and C_2 intersect at P and Q. Discuss the nature and number of solutions for each position of L relative to these two circles, as in Exercises 32-36.

39. As in Exercise 38, but suppose that C_1 and C_2 are externally tangent at T.

(10)   CCC

This is the last problem of this set and is often called by itself the "Problem of Apollonius" or the "Circle of Apollonius." The given circles may lie in various relative positions, any one of which may lead to numbers of solutions. (Can you draw the given circles so there is *no* solution?)

We discuss the most general case in which the circles are exterior to each other and which leads, in general, to eight solutions. We draw only one — the circle which is externally tangent to all three of the given circles — but other solutions would be externally tangent to some of the given circles and internally tangent to the others. Suppose the solution were available, as usual. Circle S with radius r is to be drawn tangent to given circles C_1, C_2, C_3, with centers O_1, O_2, O_3 and radii r_1, r_2, r_3. (See the figure on page 72.)

With the previous solution in mind, we can "expand" the solution S, concentrically to S', whose radius would then be $r + r_1$. Then we "shrink" circle C_1 to its center O_1, circle C_2 to circle C'_2 whose radius would then be $r_2 - r_1$, and circle C_3 to circle C'_3 whose radius would then be $r_3 - r_1$. Thus, circle S' would pass through O_1 and be tangent to circles C'_2 and C'_3, which puts us neatly back to problem 8 (PCC), since we already have point O_1 and can easily construct circles C'_2 and C'_3. Of course, we easily get the solution S by "shrinking" S' concentrically, with center O and radius $OT = OO_1 - r_1$.

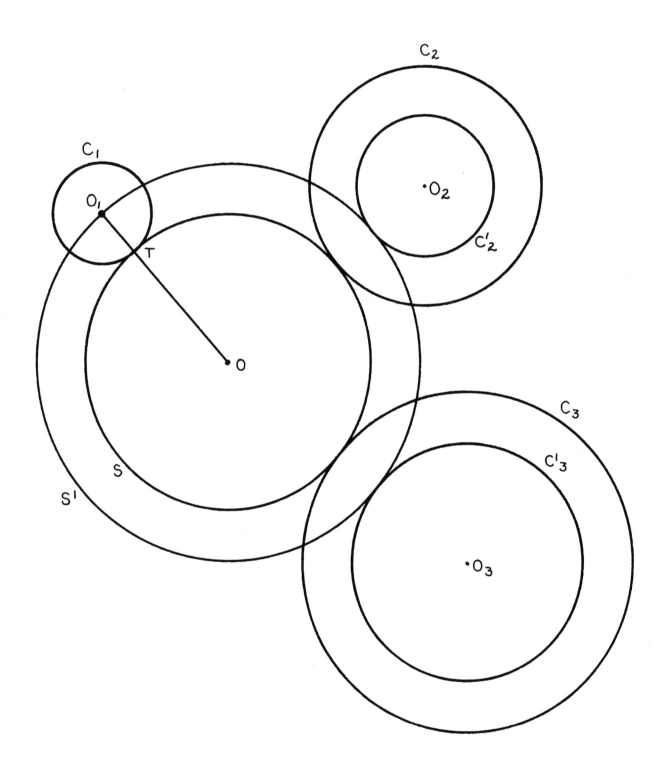

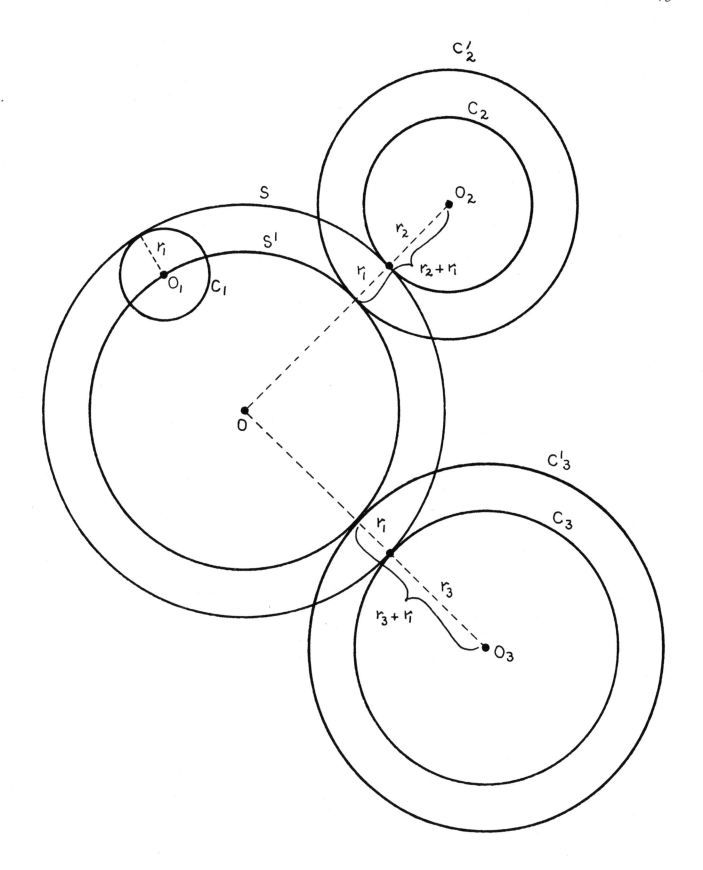

We've sketched another solution, in which S is internally tangent to circle C_1, but externally tangent to circles C_2 and C_3, as shown in the figure on page 73. The solution goes very much as before, with circle S' still found as a solution to PCC, but in this case, circle C_2' has radius $r_2 + r_1$ and circle C_3' has radius $r_3 + r_1$, as indicated.

This finishes our discussion of all ten problems, but you would be missing a lot of interesting geometry if you did not try to clear up all those delightful special cases that arise when we move the given parts around a bit, as in the exercises.

Exercises

We indicate, as usual, special situations which may exist among the given circles, which lead to different solutions and different numbers of solutions. You are asked to draw the figures and to discuss and carry out the constructive solutions.

40. C_1 is inside C_2 (not tangent), which is inside C_3 (not tangent).
41. C_1 is inside C_2 (not tangent), which is internally tangent to C_3.
42. C_1 is inside C_2 (not tangent), and C_3 intersects both C_1 and C_2.
43. C_1 is inside C_2 (not tangent), and C_3 intersects C_2, but not C_1.
44. C_1 is inside C_2 (not tangent), and C_3 intersects C_2 and is tangent to C_1.
45. C_1 is inside C_2 (not tangent), and C_3 is tangent to both C_1 and C_2.
46. C_1 is inside C_2 (not tangent), and C_3 intersects C_1, but not C_2.
47. C_1 is inside C_2 (not tangent), and C_3 is externally tangent to C_2.
48. C_1 is inside C_2 (not tangent), and C_3 is also inside C_2 (not tangent), but outside C_1.
49. C_1 is inside C_2 (not tangent) and C_3 is also inside C_2 (not tangent) and C_1 and C_3 are externally tangent.
50. C_1 is internally tangent to C_2 at T_1, and C_3 is tangent to both at T_1 (4 cases).
51. C_1 is internally tangent to C_2 at T_1, and C_3 is tangent to both, but not at T_1.
52. C_1 is internally tangent to C_2 at T_1, and C_3 intersects both circles at T_1 and at other points.
53. C_1 is internally tangent to C_2 at T_1, and C_3 intersects both circles, but not at T_1.
54. C_1 is internally tangent to C_2 at T_1, and C_2 is internally tangent to C_3 at T_2.
55. C_1 is internally tangent to C_2 at T_1, and C_3 is internally tangent to C_2 not at T_1 and intersects C_1.
56. C_1 is internally tangent to C_2 at T_1, and C_3 is tangent to C_1 and intersects C_2.
57. C_1 is internally tangent to C_2 at T_1, and C_3 is externally tangent to C_2 at T_2.
58. C_1 is internally tangent to C_2 at T_1, and C_3 is exterior to them both.
59. C_1 intersects C_2 at P and Q. Discuss the possible positions of C_3 relative to the given circles and the consequent solutions.

Chapter V

CONSTRUCTIONS WITH RESTRICTIONS ON TOOLS

To this point, we have followed the suggestion of Plato by considering geometric constructions valid when our tools were limited to the unmarked straightedge and a pair of compasses. In Chapter I, we saw that by augmenting our arsenal of tools, previous "impossible" construction became possible. For example, when we considered the marked "straightedge," the previously impossible angle trisection became constructible. Wouldn't you now think that if we reduced our arsenal of construction tools, fewer constructions would then be possible? Much to the surprise of many, this is not always so! In this chapter, we will place certain restrictions on our construction tools in a variety of ways, and in each case, we will show that our range of possible constructions is left intact. Let's begin by considering the problem which arises when we eliminate the straightedge from our arsenal and attempt to perform geometric constructions, using the compasses alone.

5.1 Constructions with Compasses Alone (Mascheroni Construction).

When faced with the prospect of performing geometric constructions without a straightedge, you may wonder how a straight line can be constructed. Quite obviously, a straight line cannot actually be constructed. However, for our purposes, whenever two points are determined, we shall *imagine* the straight line joining them. Later in this section, we will develop a method for placing as many points as we choose on such an "imagined" straight line. This may in part satisfy those who will feel more comfortable when more than two points of a line are visible.

For many years, it was felt that the compasses were a more useful instrument than the straightedge. People could not be assured that a perfect straightedge could be created. Skeptics always felt that the edge inevitably had some flaws, no matter how carefully it was fabricated. A straight line could only be copied as accurately as any existing straightedge, whereas a circle is created anew each time. For years, people were performing certain constructions with compasses only. For example, such constructions as determining the vertices of an equilateral triangle or dividing a circle into six congruent arcs were commonly done by using compasses alone. Hence, by 1797, the stage was certainly set for Lorenzo Mascheroni (1750-1800), a professor of mathematics at the University of Pavia, Italy, to present his popular work, *Geometry of the Compasses.* In this famous book, Mascheroni proved that all constructions which formerly required both straightedge and compasses may actually be done by using compasses alone. (We understand, of course, that though we cannot actually draw the line determined by two given points, we can find as many points of it as we please.) Consequently, geometric constructions using compasses alone are called "Mascheroni Constructions."

For a while in 1928, mathematicians felt a bit awkward referring to the compasses constructions as "Mascheroni Constructions." In that year, a Danish mathematician discovered a book written in 1672 by a countryman of his, Georg Mohr (1640-1697), a rather obscure mathematician. Mohr's book contains arguments similar to those of Mascheroni. Since we feel that Mascheroni did arrive at his conclusions independently, his name is used to identify these compasses constructions.

Before proving that compasses alone can replace straightedge and compasses for all possible constructions, let's inspect a few of the simpler constructions, using compasses alone.[*]

Mascheroni Construction 1:

Find the point E on $\overrightarrow{AB}$ so that AE = 2(AB). Consider the line segment $\overline{AB}$.

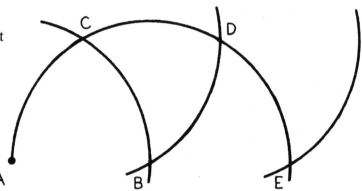

Draw arc (B, AB). The let arc (A, AB) intersect arc (B, AB) at C, let arc (C, AB) intersect arc (B, AB) at D, and let arc (D, AB) intersect arc (B, AB) at E.

Certainly, AB = BE. (Why?) Yet how can we be sure that point E is on $\overrightarrow{AB}$? Suppose we investigate this problem. What type of triangle is $\triangle$ ABC? $\triangle$ CBD? $\triangle$ DBE? . . . The answers to these questions also lead us to the measures of $\angle$ ABC, $\angle$ CBD and $\angle$ DBE, which are each 60°.

Exercise

1. Prove that point E, above, must be on $\overrightarrow{AB}$.

Now that we have discovered a method for constructing a line segment whose measure is double that of the given line segment, can you develop a method for constructing a line segment which is triple, quadruple or quintuple the measure of the given line segment? How about merely repeating our original method?

Mascheroni Construction 2:

Construct a line segment whose measure is n times the measure of a given line segment, where n = 1, 2, 3, 4, . . .

From the previous construction, we have established that AE = 2(AB). In the figure at the top of page 77, we will begin with the construction as in the figure above, then continue by letting (E, AB)

---

[*] For further exploration of this topic, see A. Kostovskii, *Geometrical Constructions with Compasses Only*, (Moscow, USSR: Mir Publishers, 1986), available in the United States from Dale Seymour Publications.

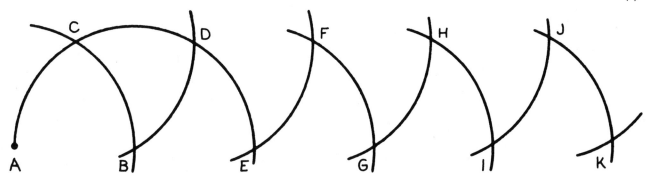

intersect (D, AB) at F; (F, AB) intersect (E, AB) at G; (G, AB) intersect (F, AB) at H; (H, AB) intersect (G, AB) at I; (I, AB) intersect (H, AB) at J; and (J, AB) intersect (I, AB) at K.

Why are the points A, B, E, G, I and K collinear? What would you surmise is true about the measure of $\overline{AB}$, $\overline{BE}$, $\overline{EG}$, $\overline{GI}$ and $\overline{IK}$? Can you now determine the segment whose measure is 3(AB)? 4(AB)? 5(AB)? We have therefore satisfied the original problem of constructing a line segment with measure n(AB), where n = 1, 2, 3, 4, ... and where AB is given.

Exercise

2. Prove that the preceding construction does what it purports to do.

So far, we have been able to locate additional points on a line. Let's now find some more points on our line $\overleftrightarrow{AB}$.

Mascheroni Construction 3:

Construct a line segment whose measure is $\frac{1}{n}$ th the measure of a given line segment.

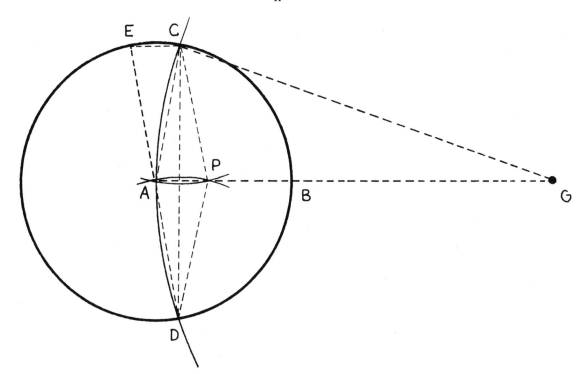

To illustrate this construction, we will merely choose any value for n, say 3. Let us again consider the line segment $\overline{AB}$. We have elected to construct a line segment whose length will be 1/3rd that of $\overline{AB}$. It might be wise to keep in mind throughout this construction that n may have just as easily taken on values 2, 4, 5, 6, 7, . . .

To begin our construction, we will construct AG = 3AB exactly as we did in Mascheroni Construction 2. Draw circle (A, AB). Then let the arc (G, GA) intersect the circle (A, AB) at points C and D. The point P, the intersection point of arcs (C, CA) and (D, DA), is a trisection point of $\overline{AB}$. That is, $AP = \frac{1}{3} AB$.

Exercises

3. How can you locate the other trisection point of $\overline{AB}$ without repeating this construction?

4. Perform the construction called for in exercise 3.

It now seems appropriate that we verify this construction. Remember, the dashed lines in the preceding illustration are merely there to help us in the proof; we had no need of them in the actual construction. We must first prove that P lies on $\overleftrightarrow{ABG}$. Since points A, P and G lie on the perpendicular bisector of $\overline{CD}$ (why?), A, P and G are collinear. Isosceles $\triangle CGA \sim$ isosceles $\triangle PCA$ (why?). Therefore, $\frac{AP}{AC} = \frac{AC}{AG}$. Since AC = AB, $\frac{AP}{AB} = \frac{AB}{AG}$. But AG = 3(AB) or $\frac{AB}{AG} = \frac{1}{3}$, therefore: $\frac{AP}{AB} = \frac{1}{3}$. Hence, $AP = \frac{1}{3}(AB)$.

Exercise

5. Carry through and prove an analogous construction for n = 5; 7.

An interesting alternate method for locating P may be employed. Using Mascheroni Construction 1, we may easily find the point E diametrically opposite point D; that is, $\overline{DAE}$ is a diameter of circle (A, AB). Since ECPA (from the figure in Mascheroni Construction 3) is a parallelogram (why?), EC = AP. Therefore, we may locate point P in an alternate way by getting the intersection of arc (A, EC) and arc (C, CA).

Exercises

6. Prove that quadrilateral ECPA is a parallelogram.

7. Perform the construction described above and compare it to the previous method, using n = 5; n = 7.

Mascheroni Construction 4:

Through point P (not on $\overleftrightarrow{AB}$) construct the line perpendicular to $\overleftrightarrow{AB}$.

This construction is simply done by drawing arc (A, AP) and arc (B, BP), which also intersect in point Q. The line $\overleftrightarrow{PQ} \perp \overleftrightarrow{AB}$.

To prove the validity of this construction, we notice that points A and B are each equidistant from P and Q, and thereby determine the perpendicular bisector of $\overline{PQ}$. How does this construction compare with Construction 7 in Chapter I?

Now that we have demonstrated that some geometric constructions are possible using only compasses, let's verify Lorenzo Mascheroni's conjecture that all constructions possible with the usual tools, straightedge and compasses, are also possible using only compasses.

How shall we go about proving this conjecture? Must we necessarily perform *all* possible constructions, using only compasses, which we are able to do using both straightedge and compasses? Although this would certainly be an acceptable procedure, it would be impossible to complete, since there are an infinite number of such constructions. Part of the beauty of mathematics is that we usually find elegant schemes to replace seemingly arduous tasks. For example, when proving triangles congruent, we do not have to prove all corresponding sides and angles congruent; we merely employ one of the familiar congruence theorems. Here, too, we shall not prove Mascheroni's conjecture by demonstrating its validity for all constructions; rather, we will attempt to discover a short cut.

What are the fundamental constructions upon which all other constructions are dependent? Following is a list of the fundamental constructions. We shall assume that any construction using both straightedge and compasses is merely a finite number of successions of these.

1. Draw a straight line through two given points.
2. Draw a circle with a given center and a given radius.
3. Locate the points of intersection of two given circles.
4. Locate the points of intersection of a straight line (given by two points) and a given circle.
5. Locate the point of intersection of two straight lines (each of which is given by two points).

Although we cannot completely satisfy the first construction on the above list, we have demonstrated in Mascheroni Construction 1 and 2 that we are able to place additional points on a line given by two points. Essentially, we are able to get all the crucial points on a line.

The second and third constructions on the above list are done directly with compasses only, and therefore need no further discussion.

Suppose we consider now the fourth construction on the list.

Mascheroni Construction 5:

Locate the points of intersection of a straight line given by two points, A and B, and a given circle, (O, r). What are the two cases to be considered here?

Case 1:

The center of the given circle does not lie on the given line. Therefore, consider circle (O, r) and the straight line $\overleftrightarrow{AB}$ shown in the figure at the right.

Find Q, the point of intersection of arcs (B, BO) and (A, AO). Then draw the circle (Q, r). The points of intersection of circles (O, r) and (Q, r), P and R, are the required points of intersection of $\overleftrightarrow{AB}$ and circle (O, r).

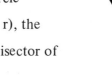

The proof of this construction is rather simple. Point Q was chosen so as to make $\overleftrightarrow{AB}$ the perpendicular bisector of $\overline{OQ}$. (Why?) By drawing a circle (Q, r) congruent to and intersecting circle (O, r), the common chord $\overline{PR}$ is also the perpendicular bisector of $\overline{OQ}$.

Exercises

8. Prove that $\overleftrightarrow{PR}$ is the perpendicular bisector of $\overline{OQ}$.

9. How does this (Exercise 8) prove that A, P, R and B are collinear?

10. State the conclusion of this proof.

Case 2:

The center of the given circle lies on the given line. Therefore, consider the circle (O, r) and the straight line $\overleftrightarrow{AB}$ shown in the figure at the right.

Draw a circle (A, x), where x is large enough to intersect (O, r) in two points, S and T. The midpoints P and R of both major and minor arc $\overset{\frown}{ST}$, will be the required points of intersection of circle (O, r) and $\overleftrightarrow{AB}$.

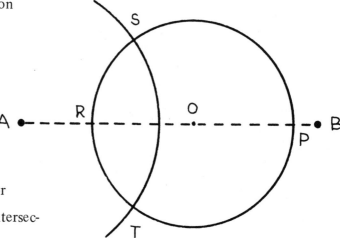

Exercise

11. Complete the preceding construction and prove that it does what it purports to do. (See below.)

Our only remaining problem here is to develop the construction for bisecting a given arc. Let's consider this problem as a separate construction.

Mascheroni Construction 6:

Bisect a given arc $\overset{\frown}{ST}$.

To begin our construction, we let OS = OT = r, where O is the center of the circle of which $\overset{\frown}{ST}$ is an arc. Let chord ST = d. Draw the circle (O, d). Then draw circles (S, SO) and (T, TO) intersecting circle (O, d) at points M and N, respectively. Next draw arcs (M, MT) and (N, NS) which meet at point K. By drawing arcs (M, OK) and (N, OK), their intersections will be our desired bisection points C and D.

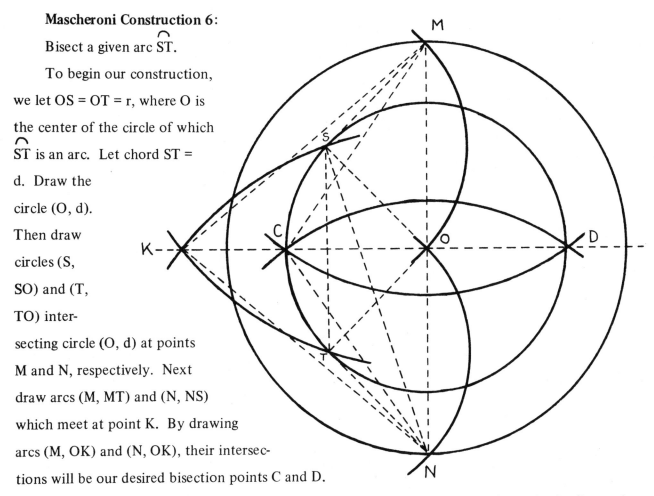

To facilitate the proof of this construction, we will use the dashed lines shown in the figure above. Quadrilaterals SONT and TOMS are parallelograms, since both pairs of opposite sides in each quadrilateral are congruent. Why may we then conclude that M, O and N are collinear? Since CN = CM and KN = KM, $\overleftrightarrow{KC} \perp \overline{MN}$ at O. Why? Why is $\overleftrightarrow{CO} \perp \overleftrightarrow{ST}$? Having confirmed this, $\overleftrightarrow{CO}$ bisects $\overline{ST}$ and $\overset{\frown}{ST}$. Why? Where would you want point C to lie with respect to circle (O, r)? If we could show that C is on circle (O, r), our proof would be complete. One way of doing this is to show that CO = r.

$$(SN)^2 + (TO)^2 = 2(SO)^2 + 2(ST)^{2*}$$
$$(SN)^2 + r^2 = 2r^2 + 2d^2$$
$$(SN)^2 = r^2 + 2d^2 \quad (I)$$

* For a proof of this statement, see Alfred S. Posamentier and Charles T. Salkind, *Challenging Problems in Geometry* (Palo Alto, Calif.: Dale Seymour Publications, 1988), p. 217.

From right $\triangle$ KON we obtain the following: $(KN)^2 = (NO)^2 + (KO)^2$. But KN = SN, therefore:

$$(SN)^2 = (NO)^2 + (KO)^2 = d^2 + (KO)^2 \text{ (II)}$$

Therefore, from (I) and (II):

$$r^2 + 2d^2 = d^2 + (KO)^2$$

or: $$(KO)^2 = r^2 + d^2$$

Lastly, from right $\triangle$ CON, we obtain the following:

$$(CO)^2 + (NO)^2 = (CN)^2$$

or: $$(CO)^2 = (CN)^2 - (NO)^2 = (KO)^2 - d^2 = r^2 + d^2 - d^2 = r^2.$$

Hence, CO = r, and our proof is complete.

Our remaining task is to show that the fifth construction on our original list can be done with compasses only.

Mascheroni Construction 7:

Locate the point of intersection of two straight lines each of which is given by two points, $\overset{\leftrightarrow}{AB}$ and $\overset{\leftrightarrow}{CD}$.

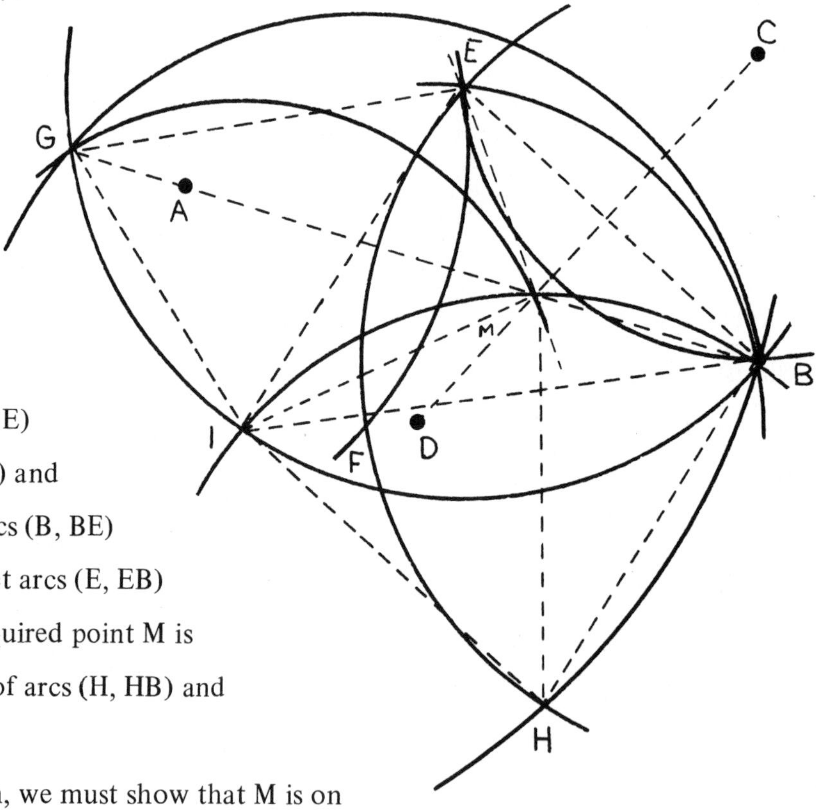

To begin our construction, let arcs (C, CB) and (D, DB) meet at E. Similarly, let arcs (A, AE) and (B, BE) meet at F. Also let arcs (E, EB) and (F, FB) meet at G. Then let arcs (B, BE) and (G, GB) meet at H. Also let arcs (E, EB) and (H, HB) meet at I. Our required point M is then found as the intersection of arcs (H, HB) and (I, IG).

To prove this construction, we must show that M is on $\overset{\leftrightarrow}{AB}$ and $\overset{\leftrightarrow}{CD}$.

Can you explain why EI = EB = BH = HI? Also IM = IG; therefore, $\overset{\frown}{IM} \cong \overset{\frown}{IG}$. Why? Since m $\angle$ IBM = ½m$\overset{\frown}{IM}$ and m $\angle$ IBG = ½m$\overset{\frown}{IG}$, we conclude that m $\angle$ IBM $\doteq$ m $\angle$ IBG. Explain why M must therefore lie on $\overset{\leftrightarrow}{BG}$. $\overset{\leftrightarrow}{AB}$ and $\overset{\leftrightarrow}{BG}$ are each the perpendicular bisector of $\overline{EF}$. Why? Now explain why M must therefore lie on $\overset{\leftrightarrow}{AB}$.

We must now show that M lies on $\overleftrightarrow{CD}$. Why is isosceles $\triangle$ BGH $\sim$ isosceles $\triangle$ BHM? It then follows that $\dfrac{BG}{BH} = \dfrac{BH}{BM}$, but since BH = BE, we get the following proportion: $\dfrac{BG}{BE} = \dfrac{BE}{BM}$, $\therefore$ $\triangle$ GEB $\sim$ $\triangle$ EMB, since both triangles share $\angle$ MBE and the sides including this angle are in proportion. Since $\triangle$GEB is isosceles, $\triangle$ EMB must also be isosceles, that is, EM = MB. CM is therefore the perpendicular bisector of $\overline{EB}$. Why may we then conclude that M must lie on $\overleftrightarrow{CD}$? We have therefore proven that M is the intersection of $\overleftrightarrow{AB}$ and $\overleftrightarrow{CD}$.

Exercises

12. Using only compasses, construct a regular hexagon with a line segment given by two points as a side.

13. Choose two points which will determine a line segment of convenient length. Using only compasses, construct a line segment with a measure seven times that of the given line segment.

14. Bisect a given line segment given by two points using only compasses.

15. Given a circle (with its center and an external point, construct a line from the external point tangent to the given circle using only compasses. (Could you dispense with the given center?)

16. Construct the perpendicular bisector of a line segment given by two points. Use compasses only.

17. Through one of the endpoints of a line segment given by its two endpoints, construct a line perpendicular to the given line segment, using only compasses.

18. Using only compasses, divide a line segment, given by two points, into five congruent segments.

19. Find as many additional points as you can on a line given by two points. Use compasses only.

Each of the following describes a possible Mascheroni construction for the problem presented. Carry out the construction indicated by following the given instructions. Then prove that the construction does what it purports to do.

20. Construct a line parallel to a given line through a point not on the given line.
    Directions for construction: Let $\overleftrightarrow{AB}$ be the given line and P the given point not on $\overleftrightarrow{AB}$. $\overleftrightarrow{PR}$ is the required line where R is the intersection of circles (B, AP) and (P, AB).

21. Construct a regular pentagon.

Directions for construction: Draw any circle with center O. Choose any point A on the circle. Let arc (A, AO) meet circle O at B. Let arc (B, BO) meet the circle at C. Then arc (C, CO) will meet the circle at D, the point on circle O which is diametrically opposite A. Locate M, the midpoint of arc $\overset{\frown}{BC}$ (see Mascheroni Construction 6). Locate the point P, the intersection of arcs (A, AC) and (D, AC). Why should (A, OP) meet circle O at M, the midpoint of $\overline{BC}$? Let arc (M, MO) intersect circle O at E (near A), and F (near D). Let X be the intersection of arcs (E, OP) and (F, OP) which is separated from P by point O. Then $\overline{AX}$ is the side of a regular pentagon inscribed in circle O. Hence, to find all five vertices of the pentagon, we mark off AX five times along the circumference of circle O.

5-2. Constructions Without Compasses.

Having completed our discussion about geometric constructions without a straightedge, it's natural to ask if all the usual constructions can be done without using a pair of compasses. This question was fully answered by Jacob Steiner (1796-1863), a Swiss mathematician, in a book entitled *Geometrical Constructions with a Ruler, Given a Fixed Circle and Its Center*, published in 1833. Although this was not the first time this idea was stated (its first appearance was in 1822 by Jean-Victor Poncelet [1788-1867]), Steiner's book presented the first complete and systematic treatment of the proof of this idea.

As the title of Steiner's book suggests, all constructions possible with both straightedge and compasses are not possible with a straightedge alone.

In order to prove that the straightedge and a fixed circle with its center given can replace the straightedge and compasses in performing geometric constructions, we have to prove that the five fundamental constructions listed on page 79 can be done with these new methods (tools). The first and last entries on this list can easily be done without compasses. However, the remaining three entries require the aid of a fixed circle and its center. It is understood, of course, that with the restricted tools we cannot actually construct a new circle with a radius different from the given one. We can, however, find as many points as we please of any such new circle. We shall not prove that these three constructions can be done by replacing a pair of compasses with a fixed circle and its center, for the proofs would not only be lengthy but also require some advanced concepts of geometry (projective geometry).

We refer the reader to some sources where these proofs can be found.

1. Jacob Steiner, *Geometrical Constructions with a Ruler, Given a Fixed Circle and Its Center.* (Translated by M. E. Stark, *Scripta Mathematica*, New York, 1950.)

2. A. S. Smogoryhevskii, *The Ruler in Geometrical Constructions.* (Translated by H. Moss, Blaisdell Publishing Company, New York, 1961.)

3. Howard Eves, *A Survey of Geometry,* Boston: Allyn & Bacon, Inc., 1972, Volume I, pages 204-209.

We shall now present some constructions without compasses, not only to familiarize the reader with this type of construction, but also to prepare those who plan to read the proof in any of the previously-mentioned books.

Construction I (without compasses):

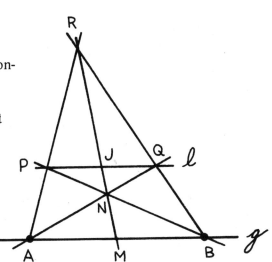

Given $\overline{AB}$, with midpoint M and point P not on $\overleftrightarrow{AB}$, construct a line containing P and parallel to $\overleftrightarrow{AB}$.

Using a straightedge, draw $\overrightarrow{AP}$. Then choose any point R on $\overrightarrow{AP}$. Then, using the straightedge, draw $\overleftrightarrow{BP}$ and $\overleftrightarrow{RM}$. Call their point of intersection N. Draw $\overleftrightarrow{AN}$ and call its point of intersection with $\overleftrightarrow{RB}$ point Q. Draw $\overleftrightarrow{PQ}$. $\overleftrightarrow{PQ} \,//\, \overleftrightarrow{AB}$.

Exercise

22. Try the construction with R on $\overline{AP}$. What happens when R is the midpoint of $\overline{AP}$?

Although in the figure above, $\overleftrightarrow{PQ}$ appears to be parallel to $\overleftrightarrow{AB}$, we must still prove that it is actually so; that is, we must prove that if N is any point on median $\overline{RM}$, of $\triangle ARB$, and $\overrightarrow{AN}$ and $\overrightarrow{BN}$ intersect $\overleftrightarrow{RB}$ and $\overleftrightarrow{RA}$ at points Q and P, respectively, then $\overleftrightarrow{PQ} \,//\, \overleftrightarrow{AB}$.

Proof: Consider the point K on $\overleftrightarrow{RNM}$, where $\overline{NM} \cong \overline{KM}$. Then draw $\overline{KA}$ and $\overline{KB}$.

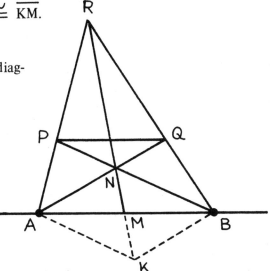

Quadrilateral ANBK is a parallelogram because the diagonals bisect each other. Therefore: $\overline{BK} \,//\, \overline{ANQ}$, and $\dfrac{RQ}{QB} = \dfrac{RN}{NK}$. Similarly, $\overline{AK} \,//\, \overline{BNP}$ and $\dfrac{RP}{PA} = \dfrac{RN}{NK}$. Thus, $\dfrac{RQ}{QB} = \dfrac{RP}{PA}$ and in $\triangle ARB$, $\overleftrightarrow{PQ} \,//\, \overleftrightarrow{AB}$.

Construction 2 (without compasses):

Given $\overleftrightarrow{AB}$ parallel to any line, ℓ ; locate the midpoint of $\overline{AB}$. (See the figure at the top of page 86.)

Select a point R not contained in $\overleftrightarrow{AB}$ nor line ℓ. Draw $\overleftrightarrow{AR}$ and $\overleftrightarrow{BR}$, which intersect line ℓ at points P and Q, respectively. Draw $\overline{AQ}$ and $\overline{BP}$, which intersect at point N. Draw $\overrightarrow{RN}$, which intersects $\overleftrightarrow{AB}$ at M. M is the midpoint of $\overline{AB}$.

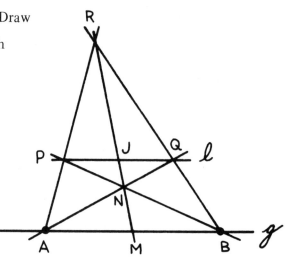

Once again, we must prove that this is actually so.

Proof: Since $\overleftrightarrow{PQ} \parallel \overleftrightarrow{AB}$, $\dfrac{PJ}{AM} = \dfrac{RJ}{RM}$ and $\dfrac{QJ}{BM} = \dfrac{RJ}{RM}$.

Therefore: $\dfrac{PJ}{AM} = \dfrac{QJ}{BM}$ or $PJ = \dfrac{QJ \cdot AM}{BM}$. Since

$\triangle PJN \sim \triangle BMN$, $\dfrac{PJ}{BM} = \dfrac{JN}{MN}$; also, since

$\triangle QJN \sim \triangle AMN$, $\dfrac{QJ}{AM} = \dfrac{JN}{MN}$. Therefore:

$\dfrac{PJ}{BM} = \dfrac{QJ}{AM}$ or $PJ = \dfrac{QJ \cdot BM}{AM}$ thus $\dfrac{QJ \cdot AM}{BM} = \dfrac{QJ \cdot BM}{AM}$ or $AM^2 = BM^2$. Hence, $AM = BM$, and M is the midpoint of $\overline{AB}$.

Exercise

23. Try this construction with R between $\overleftrightarrow{AB}$ and ℓ. What happens if R is exactly midway between them? Discuss this.

Construction 3 (without compasses):

Construct a line parallel to two given parallel lines ℓ and m, containing a point P which is not on line ℓ nor line m.

This construction is a direct consequence of the two preceding constructions. All we need to do is to bisect any segment on line ℓ . (Use Construction 2, without compasses.) Then construct the required line parallel to line ℓ . (Use Construction 1, without compasses.)

Exercise

24. Perform Construction 3, without compasses, using the previous discussion as a guide.

Construction 4 (without compasses):

From a given point, P, in the exterior region of a given circle (Q, AQ), construct a line perpendicular to diameter $\overline{AQB}$.

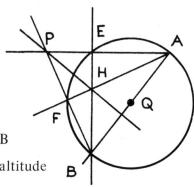

Draw $\overleftrightarrow{AP}$ and $\overleftrightarrow{BP}$, which intersect the circle at points E and F respectively. Draw $\overleftrightarrow{AF}$ and $\overleftrightarrow{BE}$. Call their point of intersection H. Draw $\overleftrightarrow{PH}$. Then $\overleftrightarrow{PH} \perp \overleftrightarrow{AB}$, since $\overline{AF}$ and $\overline{BE}$ are altitudes of $\triangle APB$ and the altitudes of a triangle are concurrent. $\overleftrightarrow{PH}$ contains the third altitude of $\triangle APB$ and hence is perpendicular to $\overleftrightarrow{AB}$.

Exercises

25. How will this construction change if P is on tangent $\overleftrightarrow{PB}$?

26. How will this construction change if P is on $\overleftrightarrow{AB}$?

27. How will this construction change if P is so situated that the required perpendicular will not intersect $\overline{AB}$ as in the figure shown here? Is the same construction possible?

In the previous construction, a circle with its center was given in the statement of the construction problem. The circle with its center was therefore obviously an essential part of the construction.

Thus, a fixed circle and its center may serve as an additional tool in the construction. This is essentially how Jacob Steiner used a circle in his straightedge constructions. As we mentioned earlier, Steiner proved that with these tools (straightedge and a fixed circle with given center), it is possible to complete all constructions which can be done, using both straightedge and compasses. We call such constructions "Steiner constructions." Let's consider such a construction.

Steiner Construction 1:

Given line ℓ and a point P not on line ℓ, construct a line containing P and parallel to ℓ, using a straightedge and a fixed circle with center given. If line ℓ contains the center Q of the fixed circle (Q, r), then we have contained in line ℓ a segment, $\overline{AB}$, whose midpoint is determined.

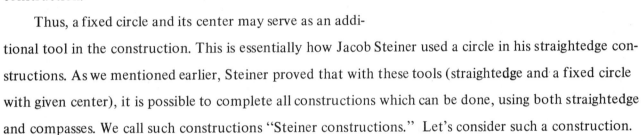

In this case, we need merely apply Construction 1 (without compasses) to obtain our desired line.

If, however, line ℓ does not contain the center Q of the fixed circle (Q, r), then we must first construct a line parallel to line ℓ. This will then permit us to use Construction 3 (without compasses). Here's where our fixed circle with given center is needed. It will enable us to construct a line parallel to line ℓ, thus allowing us to complete the required construction.

Let us now construct a line parallel to line ℓ, using the fixed circle (Q, r). (See figure at the top of page 88.)

Draw any two lines $\overleftrightarrow{CD}$ and $\overleftrightarrow{EF}$ containing center Q and not parallel to line ℓ, where points C, D, E and F are on the circle. Draw $\overleftrightarrow{CE}$ and $\overleftrightarrow{FD}$ which intersect line ℓ at points M and N, respectively. Now draw $\overleftrightarrow{MQ}$ which intersects $\overleftrightarrow{FD}$ at K, and draw $\overleftrightarrow{NQ}$ which intersects $\overleftrightarrow{CE}$ at J. $\overleftrightarrow{KJ}$ is parallel to line ℓ. Now we use Construction 3 (without compasses) to complete this construction.

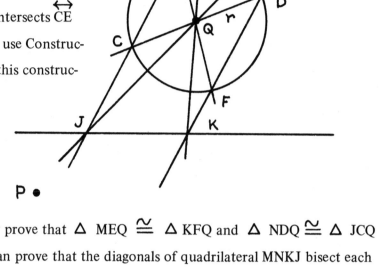

The complete proof that $\overleftrightarrow{KJ}$ // $\overleftrightarrow{MN}$ will be left as an exercise. However, we shall present a brief outline of the proof here. Since both $\triangle$ EQC and $\triangle$ FQD are isosceles and congruent, we can easily prove that $\triangle$ MEQ $\cong$ $\triangle$ KFQ and $\triangle$ NDQ $\cong$ $\triangle$ JCQ (ASA Congruence Postulate). Thus we can prove that the diagonals of quadrilateral MNKJ bisect each other. Hence, $\overleftrightarrow{KJ}$ // $\overleftrightarrow{MN}$.

Exercises

28. How will Steiner Construction 1 be affected if point P is in the interior of the fixed circle?

29. How will Steiner Construction 1 be affected if line ℓ is tangent to the fixed circle?

30. How will Steiner Construction 1 be affected if line ℓ contains interior points of fixed circle but not its center?

31. Complete Steiner Construction 1 if line ℓ contains the center of the fixed circle.

32. Complete Steiner Construction 1 if line ℓ does not contain the center of the fixed circle. Consider the various locations of line ℓ, relative to the fixed circle.

33. Complete the proof outlined in the discussion of Steiner Construction 1 that $\overleftrightarrow{KJ}$ // $\overleftrightarrow{MN}$.

Let us consider another Steiner construction.

Steiner Construction 2:

Given line ℓ and a point P not contained in line ℓ, construct a line containing P and perpendicular to ℓ, using a straightedge and a fixed circle with given center.

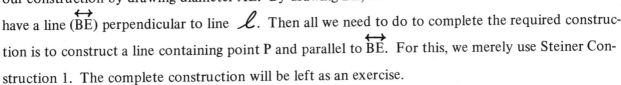

If line ℓ intersects the fixed circle (Q, QA) at points A and B and does not contain the center, we begin our construction by drawing diameter $\overline{AE}$. By drawing $\overset{\leftrightarrow}{BE}$, we have a line ($\overset{\leftrightarrow}{BE}$) perpendicular to line ℓ. Then all we need to do to complete the required construction is to construct a line containing point P and parallel to $\overset{\leftrightarrow}{BE}$. For this, we merely use Steiner Construction 1. The complete construction will be left as an exercise.

If, however, line ℓ does not intersect the fixed circle in two distinct points, then we would first have to construct a line which does intersect the fixed circle and is also parallel to line ℓ. This would then permit us to continue the construction as we did when line ℓ did intersect the fixed circle. To construct this line parallel to line ℓ, we would simply apply Steiner Construction 1. Once again we shall leave the complete construction as an exercise.

Exercises

34. Complete the construction outlined above for the case where line ℓ intersects the fixed circle in two distinct points.

35. Complete the construction outlined above for the case where line ℓ does not intersect the fixed circle.

36. How would the construction for exercise 34 change if line ℓ is tangent to the fixed circle?

37. How would the construction for exercise 34 change if line ℓ contains the center of the fixed circle?

Now that we can construct parallel and perpendicular lines, using only a straightedge and a fixed circle with center given, we can do a number of other related constructions with these tools. The following exercises will provide some applications of these new procedures. You may wish to formulate your own problems to add to this list.

Exercises

38. Construct an angle congruent to a given angle.

39. Construct a line tangent to a given circle at a given point on the circle.

40. Construct a parallelogram.

41. Construct a parallelogram similar to the parallelogram constructed in exercise 40, given one side.

42. Construct a rectangle, given its sides.

43. Use the diagram shown here as a guide to constructing the bisector of a given angle.

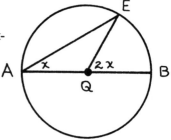

44. Use the adjoining diagram as a guide to constructing an angle whose measure is twice that of a given angle.

45. Construct the perpendicular bisector of a given line segment.

46. Locate the center of a given circle, other than the original given circle.

5-3. Constructions Using a Straightedge and Rusty Compasses (Fixed Compasses).

Suppose you were about to do some geometric constructions using a straightedge and a pair of compasses. As you began the constructions, you noticed that the compasses were stuck and could not be moved at all. Would you have to give up performing your geometric constructions until the compasses were either replaced or repaired? According to Poncelet and Steiner, you could continue your work successfully with the fixed compasses, since they would still enable you to draw a fixed circle with a given center. (This was discussed in the previous section.)

It is interesting to note that as early as the tenth century, the Arab mathematician, Abû'l - Wefâ (940-998) wrote about constructions where a pair of compasses were used with a fixed opening. Essentially, these constructions dealt with regular polygons. Albrecht Dürer, the famous German artist, included two constructions with fixed compasses constructions. The main difference between the Poncelet-Steiner constructions and the earlier fixed compasses constructions is that Poncelet and Steiner were satisfied with any fixed compasses opening and location, while their predecessors selected a convenient opening for their particular purposes. They also used more than one circle in some constructions, while Poncelet and Steiner needed only one circle.

Let us consider a few fixed compasses constructions using methods where the compasses are used more than once. Before considering the following examples, you may wish to review some rather simple examples of fixed compasses constructions which we performed as Constructions 5 and 9 in Chapter I.

The following examples of a fixed compasses construction are similar to those used by Albrecht Dürer in 1525.

Fixed Compasses Construction 1:

At point P on line ℓ , construct a line perpendicular to line ℓ .

Through point P, draw any non-perpendicular line m to line ℓ . Choose a convenient setting for the fixed compasses, say r. Draw arc (P, r) intersecting line m at Q. Then draw circle (Q, r) which intersects line ℓ at P and R. Draw $\overleftrightarrow{RQ}$ to intersect circle (Q, r) at N. Then draw $\overleftrightarrow{NP}$, our desired line, since $\overleftrightarrow{NP} \perp \overleftrightarrow{RP}$.

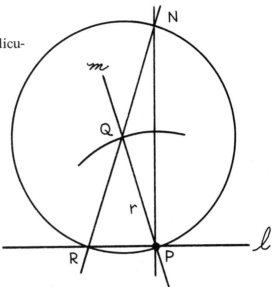

Exercises

47. Prove that the construction above actually does what we say it does.

48. Why couldn't we let line m be perpendicular to line ℓ ?

49. How does this construction differ from the Steiner constructions?

50. How could this construction be used to produce parallel lines?

51. Construct a line parallel to line ℓ and containing point N.

52. Construct a line parallel to line ℓ and not containing point N.

53. Describe how we can use Fixed Compasses Construction 1 to construct a line perpendicular to line ℓ and containing a point G which is not on line

Let us now consider a more difficult fixed compasses construction which employs some of the previously mentioned techniques. This particular construction was first published by Abû'l - Wefâ.

Fixed Compasses Construction 2:

Construct a regular pentagon, using a given segment $\overline{AB}$, contained in line ℓ , as a side.

Use Fixed Compasses Construction 1 to construct a line m perpendicular to line ℓ at point A. (We shall not show this construction in the diagram.) Draw arc (A, AB)

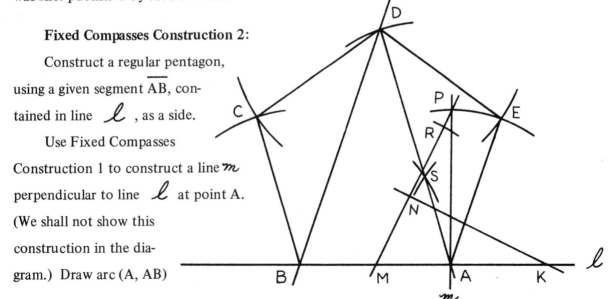

to intersect line m at P. Locate the midpoint M of $\overline{AB}$, using Construction 5 of Chapter I. Once again, for the sake of clarity, we shall not indicate the construction for locating point M in the diagram. Draw $\overleftrightarrow{MP}$. Then draw arc (M, AB) to intersect $\overline{MP}$ at R. Once again, using Construction 5 of Chapter I, construct $\overleftrightarrow{NK}$, the perpendicular bisector of $\overline{MR}$, intersecting $\overleftrightarrow{AB}$ at K. This construction is not shown in the diagram (in an effort to keep the diagram simple). Let arc (B, AB) intersect arc (K, AB) at point S. Draw $\overleftrightarrow{AS}$. Let arc (S, AB) intersect $\overrightarrow{AS}$ at D. Draw arc (A, AB) to intersect arc (D, AB) at E. Then draw arc (B, AB) to intersect arc (D, AB) at C. Draw $\overline{AE}$, $\overline{DE}$, $\overline{BC}$ and $\overline{DC}$. Polygon ABCDE is a regular pentagon.

Exercises

54. Prove that polygon ABCDE (above) is a regular pentagon.

55. Using nothing more than a straightedge and the same fixed compasses of the above construction, construct a few other non-congruent regular pentagons. Can you construct some using only the straightedge?

56. Explain how you would use Construction 9 of Chapter I to construct a regular hexagon with a side of given length, using only a straightedge and fixed compasses.

5.4. Collapsible Compasses Constructions.

In the previous section, we considered constructions performed with a straightedge and "damaged" compasses that would not permit adjustment. Let's now consider another pair of compasses which are also "damaged" but in a completely different way. These "damaged" compasses will not preserve a radius length when lifted off the plane of construction.

Before we prove that the straightedge and the collapsible compasses are actually equivalent to the straightedge and our modern compasses, that is, that any construction that can be done with one set can just as well be done with the other set, we shall perform a few simple constructions with such collapsible compasses and a straightedge.

Collapsible Compasses Construction 1:

Construct a line perpendicular to a given line ℓ at a given point P on line ℓ. Draw an arc (P, AP) intersecting line ℓ at points A and B. Then draw arc (A, AB) and arc (B, AB) to intersect at R. Draw $\overleftrightarrow{RP}$. $\overleftrightarrow{RP}$ is perpendicular to line ℓ.

57. Prove that for the preceding construction, $\overleftrightarrow{RP} \perp \overleftrightarrow{AB}$.

58. Perform the preceding construction for the case where P is not on line ℓ

59. How can we use Collapsible Compasses Construction 1 to construct a line parallel to line ℓ ?

60. Perform the construction called for in exercise 59.

Collapsible Compasses Construction 2:

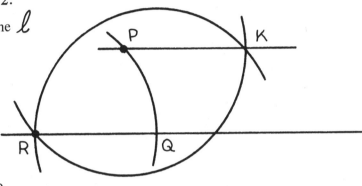

Construct a line parallel to a given line ℓ and containing a given point P.

Using any convenient point R on line ℓ , draw arc (R, RP) which intersects line ℓ at Q. Draw arc (P, RP) and arc (Q, RQ) which intersect at K. Draw $\overleftrightarrow{PK}$. $\overleftrightarrow{PK}$ is parallel to line ℓ .

61. Prove that for the above construction, $\overleftrightarrow{PK} \parallel \overleftrightarrow{RQ}$.

62. What type of quadrilateral is PKQR? Justify your answer.

63. Can this construction be performed with fixed compasses instead of the collapsible compasses used here? Why?

Our next collapsible compasses construction will provide us with a method of reproducing a given line segment anywhere in the plane of the construction (that is, with any given point as an endpoint). This will then prove that the collapsible compasses are equivalent to the modern compasses, since we will then be able to reproduce a given circle with center at any point in the plane. (Explain why!)

Collapsible Compasses Construction 3:

With a given endpoint P, construct a line segment congruent to a given line segment $\overline{AB}$. (See the figure at the top of page 94.)

Draw $\overrightarrow{AP}$. Then draw arc (P, AP) to intersect $\overrightarrow{AP}$ at C. Draw $\overline{CB}$. We must then locate the midpoint M of $\overline{CB}$. To do this, we shall use Collapsible Compasses Construction 1. That is, draw arc (C, CB) and arc (B, CB) to intersect at points E and F. $\overleftrightarrow{EF}$ is the perpendicular bisector of $\overline{CB}$. Now draw $\overleftrightarrow{PM}$. PM = ½AB. Therefore, we draw (M, PM) to intersect $\overrightarrow{PM}$ at R. Hence, $\overline{PR} \cong \overline{AB}$.

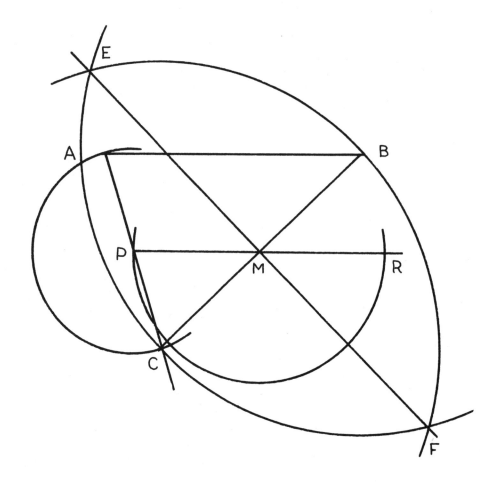

Exercises

64. Prove that for the above construction, $\overline{PR} \cong \overline{AB}$.

65. List some other things which were constructed in the previous construction.

66. Can Collapsible Compasses Construction 2 be used to replace the above construction in reproducing a given segment with a given point as endpoint? Justify your answer.

Now that we have shown that the collapsible compasses are equivalent to the modern compasses, you might want to try your hand at some constructions with a straightedge and a pair of collapsible compasses.

Exercises

For each of the following constructions, use a straightedge and collapsible compasses only.

67. Construct the bisector of a given angle.

68. Construct an angle congruent to a given angle.

69. Construct an equilateral triangle given a side.

70. Construct a regular hexagon, given a side.

71. Construct a triangle congruent to a given triangle.

Chapter VI

GEOMETRIC CONSTRUCTION: THE DOUBLE STRAIGHTEDGE

This chapter is reprinted by permission of the National Council of Teachers of Mathematics. From William Wernick, "Geometric Construction: The Double Straightedge," The Mathematics Teacher, Vol. LXIV, No. 8 (December 1971), pp. 697-704.

THE classical Euclidean constructions allow only the unmarked straightedge and a pair of compasses, now commonly called "a compass." With these limited tools we can carry through the many beautiful constructions of elementary geometry. A lack of understanding of the "rules of the game" leads to such confused and incorrect statements as "It is impossible to trisect an angle." But of course we can trisect a right angle; and as for trisecting any angle, there is a simple and elegant construction by Plato that does this very nicely, if we allow such a slight concession as two marks on the straightedge, or even one mark and the endpoint!

Suppose we are given $\angle AOB$ to trisect, and we have a "straightedge" with endpoint P and a single mark at Q. Construct a semicircle with center O and radius $\overline{OA} \cong \overline{PQ}$ and extend diameter $\overline{AOC}$ as shown in figure 1. Place the straightedge so that P lies on $\overset{\leftrightarrow}{AO}$, Q lies on the semicircle,

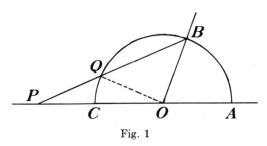

Fig. 1

and edge $\overset{\leftrightarrow}{PQ}$ goes through B on the semicircle. Then $m \angle APB = \frac{1}{3}m \angle AOB$. The proof follows easily from consideration

of the exterior angles of the two isosceles triangles, $\triangle OQP$ and $\triangle OBQ$.

There is a related construction that you may work out for yourself in which $\overline{PQ}$ is the exterior portion intercepted between C and the extension of $\overline{OB}$. Both these constructions are not acceptable for two reasons. In the first place, Euclid's original line postulate says, in effect, "If A and B are distinct points, then there is exactly one line that contains them." The corresponding construction would require that we have two distinct points of a line *before* we can draw, with a straightedge, the unique line that contains them. In both the constructions discussed we do *not* have two such points; the crucial line is determined by only one point and a placing-fitting procedure, which is not considered to be "playing the game" in accordance with the stated postulate. In the second place, both these constructions use a forbidden tool, a marked straightedge.

A careful reading of Euclid's original postulates places us under even greater restrictions than are usually imposed. "A circle may be drawn with any point A as center and passing through any other point B" does not really allow us to pick up our compasses and mark that same radius length $\overline{AB}$ from another center. The usual construction to make segment $\overline{PQ}$ congruent to a given segment $\overline{AB}$ goes something like this: "Set the compasses at radius $\overline{AB}$; then, with P as center and AB as radius . . . ." But this is not *really* playing the game, because all that the classical Euclidean circle constructions allow with our compasses is a circle with center A and radius $\overline{AB}$ or a circle with center B and radius $\overline{BA}$. When we "set the compasses at radius $\overline{AB}$" and then move them elsewhere, we are using our compasses as *dividers* that can hold this setting. If we pick up our "Euclidean compasses," we cannot hold them at a fixed opening. We may think of them, theoretically at least, as *snap-compasses*, which can keep a radius while the circle is being drawn but snap closed when they are lifted from the paper. Does this mean that the usual high school geometry construction indicated above is invalid? Not at all!

It is not difficult to show that this new restriction is more apparent than real and, in fact, that any construction possible with our usual compasses (when used as dividers that can hold a radius) is also possible with the snap-compasses (which cannot). Students, and perhaps you too, may find some interest in carrying through the familiar elementary geometric constructions using only the "pure Euclidean" tools: the unmarked straightedge and the snap-compasses.

A variant on the snap-compasses is the *rusty compass*, which can draw circles with only one given radius. With a straightedge, such a fixed compass also allows all the usual Euclidean constructions, as indicated in an excellent article by Arthur E. Hallerberg in the April 1959 issue of this journal. It is understood that we cannot actually construct a circle with a different radius, but we can find as many points of it as we please.

As a matter of fact, the Mascheroni constructions (see the fine presentation by Julius Hlavaty in the November 1957 issue of this journal) show that the compasses alone can supply all the usual constructions. It is understood here that a "line," instead of being drawn along a straightedge, is now considered determined as soon as we locate any two of its points. It is thus a significant problem to find the intersection of two of these "lines."

It is known that the straightedge alone, since it allows no interval measure, cannot perform all the Euclidean constructions. It can be used for those that involve only intersection and collinearity and is thus the essential tool for projective geometry.

I propose here a simple nonclassical tool: an unmarked strip with parallel edges—in other words, a double straightedge that is essentially, if we disregard all marks along its edges, the ordinary ruler that is usually available to all students. Since it does have an implicit measure, the fixed distance d between the parallel edges, it can do more than the straightedge alone. We shall soon see that it is mathematically equivalent to the usual tools and that with it we can perform all the usual Euclidean constructions. Of course it cannot draw a circle, but it does allow us to find as many points as we like of the circle whose center A and radius $\overline{AB}$ are given by the notation $(A, \overline{AB})$.

The following basic constructions show the use and versatility of this simple and available tool. Some proofs are omitted, since they are quite obvious, but a few are sketched briefly. Some proofs depend on concepts and properties from projective geometry or from less generally familiar parts of Euclidean geometry, but these are available in the references cited or in equivalent references.

CONSTRUCTION 1. *Construct three or more equally spaced parallel lines.*

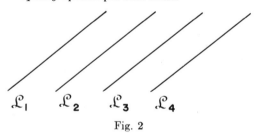

Fig. 2

Along the edges $\mathcal{E}_1$ and $\mathcal{E}_2$, draw the lines $\mathcal{L}_1$ and $\mathcal{L}_2$, which are thus parallel and d apart (fig. 2). Now place $\mathcal{E}_1$ on $\mathcal{L}_2$, and draw $\mathcal{L}_3$ along $\mathcal{E}_2$. Repeat in this way to obtain as many parallel lines as desired: $\mathcal{L}_1$, $\mathcal{L}_2$, $\mathcal{L}_3$, . . . , equally separated by the common distance d.

CONSTRUCTION 2. *Double a given segment $\overline{AB}$:*

a) When $AB \geq d$.

Place the strip so that $\mathcal{E}_1$ is on A and $\mathcal{E}_2$ is on B, then draw $\mathcal{L}_1$ and $\mathcal{L}_2$ (fig. 3). Now use Construction 1 to find $\mathcal{L}_3$, which meets line $\overleftrightarrow{AB}$ in point C. This

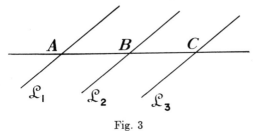

Fig. 3

construction obviously generalizes to give any integral multiple of $\overline{AB}$.

b) When $\overline{AB}$ is any given segment.

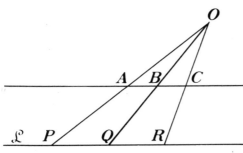

Fig. 4

Make $\mathcal{L} \parallel \overleftrightarrow{AB}$ by Construction 1, and then by Construction 2a make any two congruent segments $\overline{PQ} \cong \overline{QR}$ on $\mathcal{L}$ (fig. 4). Now, with the straightedge find $\overleftrightarrow{PA} \cap \overleftrightarrow{QB} = O$; then $\overleftrightarrow{OR} \cap \overleftrightarrow{AB} = C$. Clearly $\overline{AB} \cong \overline{BC}$, and the construction can be generalized as above.

CONSTRUCTION 3. *Bisect a given segment $\overline{AB}$:*
a)

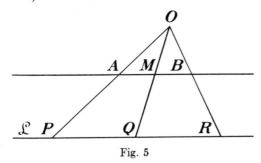

Fig. 5

As in Construction 2b find P, Q, and R

on $\mathcal{L}$ so that $\mathcal{L} \parallel \overleftrightarrow{AB}$ and $\overline{PQ} \cong \overline{QR}$, but now take $\overleftrightarrow{PA} \cap \overleftrightarrow{RB} = O$, and $\overleftrightarrow{OQ} \cap \overleftrightarrow{AB} = M$ (fig. 5). The generalizations should now be obvious: (1) divide a given segment into m equal parts; and (2) find any rational multiple, m/n, of a given segment.

b) When $\overline{AB} \geq d$.

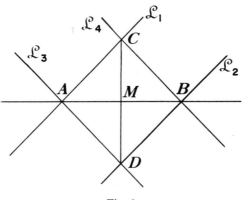

Fig. 6

For this case there is a simple construction with an extra dividend. Place the strip so that $\mathcal{E}_1$ and $\mathcal{E}_2$ lie respectively on A and B, with "positive slope," and draw $\mathcal{L}_1$ and $\mathcal{L}_2$ as shown in figure 6. Then place the strip in the symmetric "negative slope" position and draw $\mathcal{L}_3$ and $\mathcal{L}_4$ as shown. Then $\mathcal{L}_1 \cap \mathcal{L}_4 = C$; $\mathcal{L}_2 \cap \mathcal{L}_3 = D$; and $\overleftrightarrow{CD} \cap \overleftrightarrow{AB} = M$. Note that $\overleftrightarrow{CD}$ is the perpendicular bisector of $\overline{AB}$ and that $ADBC$ is a rhombus.

c) For any $\overline{AB}$.

Draw $\mathcal{L} \parallel \overleftrightarrow{AB}$ by Construction 1. From O, any point not on $\overleftrightarrow{AB}$, we have $\overleftrightarrow{OA} \cap \mathcal{L} = P$; $\overleftrightarrow{OB} \cap \mathcal{L} = Q$; $\overleftrightarrow{AQ} \cap \overleftrightarrow{BP} = R$; and $\overleftrightarrow{OR} \cap \overleftrightarrow{AB} = M$ (fig. 7). The proof that M is thus the midpoint of $\overline{AB}$ depends on the harmonic properties of a complete quadrangle as developed in almost any text on projective geometry. Briefly, the diagonals $\overleftrightarrow{PQ}$

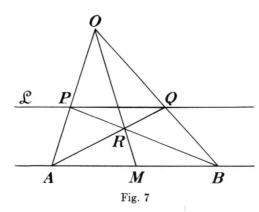

Fig. 7

and $\overleftrightarrow{OR}$ of the quadrangle $OPRQ$ divide the third diagonal, $\overleftrightarrow{AB}$, harmonically. But since $\overleftrightarrow{PQ} \parallel \overleftrightarrow{AB}$, we have $\overleftrightarrow{AB} \cap \overleftrightarrow{PQ} = \infty$; therefore (A, M, B, ∞) is a harmonic range, and thus M bisects $\overline{AB}$.

CONSTRUCTION 4. *Construct a line through a given point P and parallel to a given line $\mathcal{L}$.*

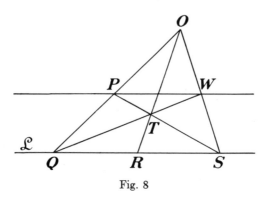

Fig. 8

On $\mathcal{L}$, make $\overline{QR} \cong \overline{RS}$ (fig. 8) by Construction 2, then on $\overleftrightarrow{QP}$ take any other point O. Then $\overleftrightarrow{OR} \cap \overleftrightarrow{PS} = T$, and $\overleftrightarrow{OS} \cap \overleftrightarrow{QT} = W$; draw $\overleftrightarrow{PW}$. The proof that $\overleftrightarrow{PW}$ is thus parallel to $\mathcal{L}$ depends on the same harmonic properties of a complete quadrangle referred to in Construction 3c.

CONSTRUCTION 5. *Construct a line through a given point P and perpendicular to a given line $\mathcal{L}$:*

a) *If P is on $\mathcal{L}$.*

By Construction 2a make $\overline{PA} \cong \overline{PB}$; then by Construction 3b draw $\overleftrightarrow{CD}$ (fig. 9). Since $\overleftrightarrow{CD}$ is the perpendicular bisector of $\overline{AB}$ it must go through P.

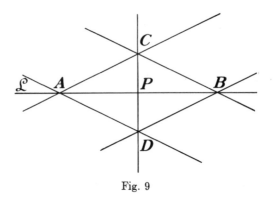

Fig. 9

b) *If P is not on $\mathcal{L}$.*

First use Construction 5a to draw $\mathfrak{M}$, any perpendicular to $\mathcal{L}$; then use Construction 4 to make a line through P and parallel to $\mathfrak{M}$, which is therefore perpendicular to $\mathcal{L}$.

CONSTRUCTION 6. *Bisect a given angle $\angle ABC$.*

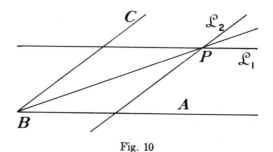

Fig. 10

By Construction 1 make $\mathcal{L}_1 \parallel \overleftrightarrow{AB}$ and $\mathcal{L}_2 \parallel \overleftrightarrow{BC}$ (fig. 10). Then $\mathcal{L}_1 \cap \mathcal{L}_2 = P$, and $\overleftrightarrow{BP}$ bisects $\angle ABC$. Clearly $\overleftrightarrow{BP}$ is a diagonal of the rhombus formed by the first steps of the construction.

CONSTRUCTION 7. *Copy a given segment $\overline{AB}$:*

a) *From point P on line $\overleftrightarrow{AB}$.*

If P is the point A or B, then this is simply Construction 2. Otherwise, draw

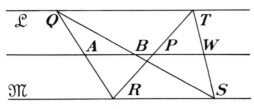

Fig. 11

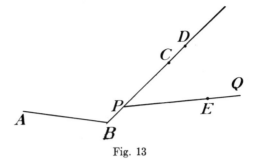

Fig. 13

$\mathcal{L}$ and $\mathfrak{M}$ $\parallel$ $\overleftrightarrow{AB}$ and on opposite sides of $\overleftrightarrow{AB}$ (fig. 11), by Construction 1. Then, from any point Q on $\mathcal{L}$ we have $\overleftrightarrow{QA} \cap \mathfrak{M} = R$; $\overleftrightarrow{QB} \cap \mathfrak{M} = S$; $\overleftrightarrow{RP} \cap \mathcal{L}$ $= T$; and $\overleftrightarrow{TS} \cap \overleftrightarrow{AB} = W$. The proof that $\overline{PW} \cong \overline{AB}$ follows readily from the fact that $\overline{AB}$ and $\overline{PW}$ are both parallel to $\overline{RS}$ and half as long as $\overline{RS}$.

b) *From A, along side $\overrightarrow{AC}$ of $\angle BAC$.*

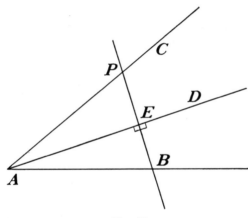

Fig. 12

First draw $\overleftrightarrow{AD}$ to bisect $\angle BAC$, by Construction 6; then draw $\overleftrightarrow{BE}$ perpendicular to $\overleftrightarrow{AD}$ (fig. 12) by Construction 5b. Then $\overleftrightarrow{BE} \cap \overleftrightarrow{AC} = P$, and $\overline{AP} \cong \overline{AB}$.

c) *From P along any line $\overleftrightarrow{PQ}$.*

First draw line $\overleftrightarrow{BP}$, then make $\overline{BC} \cong \overline{BA}$ (fig. 13) by Construction 7b. Then make $\overline{PD} \cong \overline{BC}$ by Construction 7a, and finally make $\overline{PE} \cong \overline{PD}$ by Construction 7b.

CONSTRUCTION 8. *Double a given $\angle ABC$.*

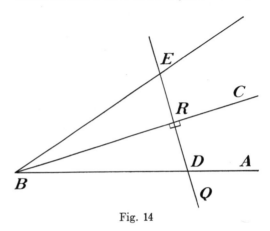

Fig. 14

From R, any point on side $\overrightarrow{BC}$, construct line $\overleftrightarrow{RQ}$ perpendicular to $\overleftrightarrow{BC}$ (fig. 14), by Construction 5a. Then $\overleftrightarrow{RQ} \cap \overleftrightarrow{AB} = D$. Double $\overline{DR}$ to $\overline{RE}$ by Construction 2, then draw $\overleftrightarrow{BE}$. Clearly $\angle CBE$ is congruent to $\angle ABC$.

CONSTRUCTION 9. *At point P on line $\mathcal{L}$ copy a given angle, $\angle ABC$.*

Obviously, if P and B coincide and if either side of $\angle ABC$ lies along $\mathcal{L}$, this is the same as Construction 8 above. We consider other situations:

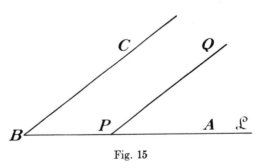

Fig. 15

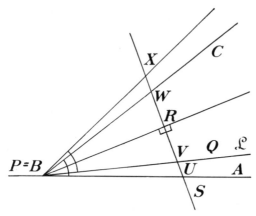

Fig. 16

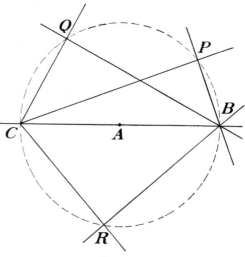

Fig. 17

$\overleftrightarrow{BC} = W$. Now, on $\overleftarrow{SRW}$ make $\overline{RX} \cong$ $\overline{RU}$, by Construction 2, then draw $\overleftrightarrow{BX}$. We can quickly show that $\angle QBX$ is congruent to $\angle ABC$, either by addition or by subtraction.

c) In all other circumstances, we have P neither on the vertex nor on either side of $\angle ABC$, and we have $\mathcal{L}$ anywhere through P.

We first draw $\overleftrightarrow{PQ} \parallel \overleftrightarrow{BA}$ and $\overleftrightarrow{PR} \parallel \overleftrightarrow{BC}$ by Construction 4, which makes $\angle QPR$ congruent to given $\angle ABC$. Now we

a) *A side of the angle, say* $\overrightarrow{AB}$*, lies on* $\mathcal{L}$.

Make $\overleftrightarrow{PQ} \parallel \overleftrightarrow{BC}$ by Construction 4 (fig. 15).

b) *P and B coincide, but neither side of the angle lies along* $\mathcal{L}$.

Let Q be any other point on $\mathcal{L}$, and then draw $\overleftrightarrow{BR}$ to bisect $\angle QBC$ (fig. 16) by Construction 6. From R, any point on $\overrightarrow{BR}$, construct $\overleftrightarrow{RS}$ perpendicular to $\overleftrightarrow{BR}$, by Construction 5a. Then $\overleftrightarrow{RS} \cap \overleftrightarrow{AB} = U$; $\overleftrightarrow{RS} \cap \mathcal{L} = V$; and $\overleftrightarrow{RS} \cap$

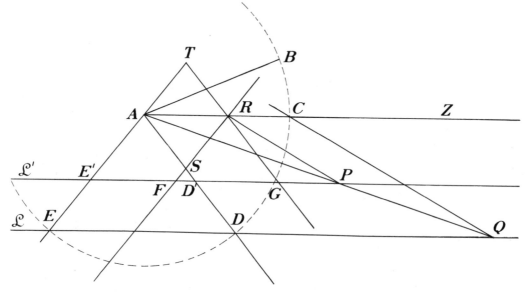

Fig. 18

can finish the job by Construction 9*b*.

CONSTRUCTION 10. *Construct any number of points of the circle with center A and radius* $\overline{AB}$, *that is, of circle* $(A, \overline{AB})$.

First we double segment $\overline{BA}$ to C, by Construction 2, then draw any number of lines through B (fig. 17). From C, we draw perpendiculars to each of these lines by Construction 5*b*. The feet of these perpendiculars are all on $(A, \overline{AB})$, since $\overline{BC}$ is a diameter of that circle.

CONSTRUCTION 11. *Find the intersection of a given line* $\mathcal{L}$ *with a given circle* $(A, \overline{AB})$.

Through A, draw $\overset{\leftrightarrow}{AZ} \parallel \mathcal{L}$ by Con-

struction 4, then make $\overline{AC} \cong \overline{AB}$ (fig. 18) by Construction 7*b*. Make $\mathcal{L}' \parallel \overset{\leftrightarrow}{AZ}$ by Construction 1 and draw any line from A to cut $\mathcal{L}'$ and $\mathcal{L}$ at P and Q. Through P draw the parallel to $\overset{\leftrightarrow}{CQ}$, by Construction 4, to cut $\overset{\leftrightarrow}{AZ}$ at R. Now, as in Construction 3*b*, construct the rhombus $ASRT$ and extend its sides to cut $\mathcal{L}'$ and $\mathcal{L}$ at E' and E and at D' and D, as shown. Then D and E are the required intersection points of $\mathcal{L}$ and circle $(A, \overline{AB})$.

Proof. $AE'/AE = AD'/AD = AP/AQ = AR/AC$. We have, from the construction, not only the rhombus $ASRT$ but also the rhombuses $AE'FR$ and $AD'GR$. Therefore, $\overline{AE'} \cong \overline{AD'} \cong \overline{AR}$; and finally, from the proportions above,

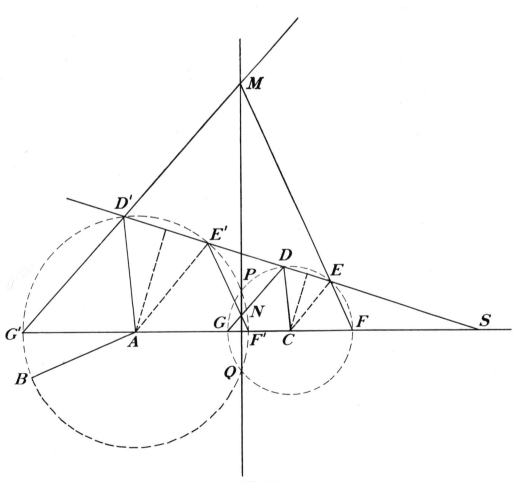

Fig. 19

$$\overline{AE} \cong \overline{AD} \cong \overline{AC} \cong \overline{AB}.$$

CONSTRUCTION 12. *Find the intersections of circles* $(A, \overline{AB})$ *and* $(C, \overline{CD})$.

This is the longest and most difficult of the constructions presented here. First we make $\overline{AD'} \cong \overline{AB}$ and parallel to $\overline{CD}$ (fig. 19), as we did in Construction 11. Then $\overleftrightarrow{D'D} \cap \overleftrightarrow{AC} = S$, which is a center of similitude for the two given circles. We then find E' and E, the other points at which the given circles cut $\overleftrightarrow{D'D}$, by first drawing perpendiculars from the centers to this line, by Construction 5a, and then copying the half-chords (Construction 2). We find the intersections of the line of centers $\overleftrightarrow{AC}$ with both circles—that is, G', F', G, and F—by Construction 7b. At this stage we have four pairs of antihomologous points—(D', E), (E', D), (G', F), and (F', G). These determine two pairs of antihomologous chords, $(\overline{G'D'}, \overline{EF})$ and $(\overline{GD}, \overline{E'F'})$, which in turn determine the points M and N, as indicated. But M and N determine the radical axis $\overleftrightarrow{MN}$ of the two given circles. Since this radical axis must contain the intersection points of the two circles, the problem has been reduced to that of finding the intersections of a line, $\overleftrightarrow{MN}$, and a circle, either $(A, \overline{AB})$ or $(C, \overline{CD})$; and this has already been done in Construction 11.

With these constructions we have now shown that the double straightedge is mathematically equivalent to the usual Euclidean tools, since with appropriate definitions it can perform the same tasks. You and your students may enjoy the specific details of similar constructions beyond the ones presented here. Would you like to try constructing an equilateral triangle with a given side, or the mean proportional to two given segments? Suppose we have, instead of the double straightedge, a single cutout triangle with definite but unspecified sides and angles. With this nonclassical tool we can also perform all the usual constructions, understanding, of course, that though we cannot construct the actual circle with given center and radius, we can find as many points of it as we please. A tool or set of tools with such capabilities may be called "adequate," and this paper has shown that the double straightedge is adequate. The assertion is made just above that the cutout triangle is also adequate. You may enjoy showing that the following sets are also adequate:

1. A straightedge and a single circle with its center

2. A cutout angle of definite but unspecified size whose sides are indefinitely long

3. Ordinary compasses and a "short straightedge"

It is hoped, finally, that a detailed examination of the uses and limitations of the several variants on the classical Euclidean tools may lead to a deeper appreciation of the structure and beauty of the original postulational system.

In addition to the references mentioned above, there may be mentioned four others:

1. *Theorie der Geometrischen Konstructionen*, by August Adler, Leipzig, 1906. This German book is an absolute gold mine of information in these matters.

2. *Geometrical Tools: A Mathematical Sketch and Model Book*, by R. C. Yates, rev. ed. 1949, Education Publishers. This is a workbook, in which you are expected to fill in construction details of a surprisingly extensive collection of interesting and challenging exercises. With the hints and suggestions, these are not too difficult; but this is not a book for quick and easy reference, though you will certainly learn a lot of geometry if you go through the exercises.

3. For the concepts and relations in projective geometry, you may consult almost any text on that subject. One that is quite satisfactory and generally available is *Projective Geometry*, by Frank Ayres, Jr., Schaum Outline Series, McGraw-Hill.

4. *College Geometry*, by Nathan Altschiller-Court, 2d ed., Barnes & Noble. This great classic has all you may want to know, and perhaps more, about radical axes, antihomologous points, and so on.